H.M.S. LONDON

From Fighting Sail to the Arctic Convoys & Beyond

Iain Ballantyne

Pen & Sword
MARITIME

**This book is for all those who have found that
just wars sometimes fall short of
high expectations.**

First published in Great Britain in 2003 by Leo Cooper
Republished in this format in 2022 by
Pen & Sword Maritime
An imprint of
Pen & Sword Books Ltd
Yorkshire – Philadelphia

ISBN 978 1 39901 286 7

A CIP catalogue record for this book is available from the British Library.

Printed and bound in the UK by CPI Group (UK) Ltd, Croydon, CR0 4YY.

Pen & Sword Books Limited incorporates the imprints of Atlas, Archaeology, Aviation, Discovery,
Family History, Fiction, History, Maritime, Military, Military Classics, Politics, Select, Transport,
True Crime, Air World, Frontline Publishing, Leo Cooper, Remember When, Seaforth Publishing, The
Praetorian Press, Wharncliffe Local History, Wharncliffe Transport, Wharncliffe True Crime, White
Owl and After the Battle.

For a complete list of Pen & Sword titles please contact

PEN & SWORD BOOKS LIMITED
47 Church Street, Barnsley, South Yorkshire, S70 2AS, England
E-mail: enquiries@pen-and-sword.co.uk
Website: www.pen-and-sword.co.uk

Or

PEN AND SWORD BOOKS
1950 Lawrence Rd, Havertown, PA 19083, USA
E-mail: Uspen-and-sword@casematepublishers.com
Website: www.penandswordbooks.com

H.M.S. LONDON

Contents

HMS *LONDON* Battle Honours 1652-1991

Kentish Knock 1652, Gabbard 1653, Scheveningen 1653, St James' Day Fight 1666, Sole Bay 1672, Schooneveld 1673, Texel 1673, Barfleur 1692, Chesapeake 1781, Île de Croix 1795, Copenhagen 1801, Marengo 1806, Crimea 1854-5, Dardanelles 1915, Atlantic 1941, Arctic 1941-43, Kuwait 1991.

A Note on Ships & Battle Honours

There is a bewildering, and sometimes conflicting, array of information on the *Londons* of the British fleet and their battle honours. The warships included in this book have been selected using the official battle honours attributed to the name HMS *London* as a guide, combined with the Royal Navy's own definition of what ships were the true bearers of the title. I have differed from some naval records regarding the *Londons* of the early seventeenth century, by considering the *London* of 1636 as the first fully fledged English warship to bear the name, rather than the vessel of 1656 (which I consider to be the third *London*).

When it comes to battle honours, Lowestoft 1665 in fact involved a hired merchant vessel called *London*. She was not the official bearer of the title HMS *London*, as that vessel was under construction at the time. Therefore the honour, Lowestoft 1665, cannot be attributed to HMS *London*.

Confusion also arises when it comes to the proper names of battles in the Anglo-Dutch naval wars. The two clashes at Schooneveld and Texel in 1673 are sometimes merged under the single battle honour, De Ruyter 1673. Some lists of *London* battle honours give alternative names for the three main Anglo-Dutch clashes of 1652-53. In this book North Foreland (September 1652) is referred to by the generally accepted title Kentish Knock, while Lowestoft (June 1653) is Gabbard and Camperdown (July 1653) is Scheveningen.

Hopefully, I have resolved these issues satisfactorily.

HMS *London*'s battle honours, as displayed on the cruiser in the late 1940s. This appears to be a selection of 'highlights', with 'Camperdown' instead of 'Scheveningen' and 'De Ruyter' used as an alternative to 'Schooneveld' and 'Texel'. *Peter Seaborn Collection.*

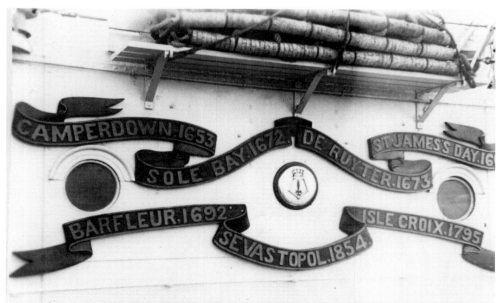

HMS *London's* ship's badge.
Image courtesy of T. Elliott.

ACKNOWLEDGEMENTS

In piecing together the story of HMS *London* I am indebted to a large cast of people who have given me an incredible amount of support as well as practical assistance. Mike Overton very generously allowed me to use material downloaded from a web site that he has created in honour of the cruiser HMS *London* (http://freespace.virgin.net/michael.overton1/hms.htm).

Mike's father, Derek, served in HMS *London* during the Yangtze Incident of 1949 and contributed via his son. I am in their debt.

Through Mike I was put in touch with the HMS *London* 1947-49 Association. Its members deserve my special thanks for loaning images from their private collections, allowing me to interview them and contributing in other ways, especially Don Chidlow, John Parker, Christopher Parker-Jervis and Neil Stewart. Frederick G. Harwood, who was decorated by King George VI with the Distinguished Service Medal for his courage and endurance under fire during the Yangtze Incident, also contributed but sadly died before publication. Another casualty of time was Benny Goodman, who made some valuable contributions to the chapters on the Second World War, but unfortunately passed away in the late summer of 2002.

Alan J. Howlett, whose brother served in *London* during the same 1947-49 commission has forwarded a number of images to me and also deserves thanks. Captain Tommy Catlow RN (Retd.) mailed me a copy of his excellent book, *A Sailor's Survival*, a much appreciated gift that considerably enhanced the texture of Chapters Fifteen and Sixteen.

Bill Jacobs provided two key images via Mike Overton who also put me in touch with others who served in HMS *London* during the Second World War. Graham Bramley, Norman Brigden, Gordon Bruty and Ted Huke all deserve my thanks for allowing me to interview them and giving material from their private collections. Brian Whitcombe whose father, Alexander, served in HMS *Cumberland* while she was in company with the *London* during the bombardment of Car Nicobar in 1944, deserves thanks for providing a dramatic image of that event. Research on the web and two adverts for contributions to the book in the British monthly naval newspaper *Navy News* yielded a flurry of responses from around the world, either via e-mail or letters. Articles in *Legion* (the magazine of the British Legion) and the *Evening Herald*, Plymouth, also sparked help from those who served in the last three *Londons*, or had some other connection with the warships. My gratitude goes to the following who responded to the adverts and articles, providing their memories and, in several cases, key photographs and original documents from their collections: David Atkins, Bob Boynton, Charlie Breach, Casper Cardwell, Douglas Clark, C. Clifford, Eddie Cowling, Charles Cox, Yorkie Cunningham, Fred Dale, Ted Davis, Bryan Dunster, Walter Eglinton, Charlie Ellis, C.W. Ellis, Malcolm Farmer, Ken Fleming, David Gilchrist, R.D. Houghton, Ian Inskip, Patrick Lambert, Warwick Luke, Mac McComb, Gordon Moores, Mike North, Brian Parker, David Parker, Douglas Parker, Gordon Perry, Terry Potton, Albert Price, John Reekie, Rod Saul, Peter Seaborn, Martin Sykes, Ken Tamon, Martin Weaver, Bill West, Harry Williams and Waldie Willing.

John Cunningham and Victor Parker – who sailed in *London* as young Army officers in the 1940s – played their part in this maritime story with their recollections and photographs.

The eyewitness accounts of serving members of the Royal Navy were invaluable in bringing the story of the last *London* to life. Despite his busy schedule as Assistant Chief of the Naval Staff, Rear Admiral Tim McClement made himself available for an interview about his time as the warship's Commanding Officer and also provided photographs and other material from his private papers. I am extremely grateful to Rear Admiral McClement for his vital input.

Lieutenant Commander Lee Abernethy was generous with both his time and material from his own private papers, while Warrant Officers Tim Allen, Bob Burton and Chris Norris made important contributions.

Commodore Richard Leaman, Director of Corporate Communications (Navy) and his staff, provided every assistance particularly in arranging approval for participation of the serving RN personnel.

Two other former Commanding Officers of the last *London* who enriched the book were Captain Doug Littlejohns RN (Retd.) and Captain James Taylor RN (Retd.). The legendary Captain Littlejohns not only agreed to be interviewed, but also provided photographs from his private collection, all of which have greatly enhanced the book.

Having become acquainted with Captain Chris Craig RN (Retd.) during his time as the CO of Britain's Gulf War naval task group, it was a pleasure to reintroduce myself by asking for his contribution to a book on his flagship. The author of an impressive book on his part in both the Falklands and Gulf conflicts – *Call For Fire* – Captain Craig was able to give an unrivalled and incisive insight into the events *London* helped to shape and also freely offered material from his private papers. As ever, Captain Craig impressed me with his humanity, just one of several excellent qualities that made him such an inspiring leader in the Gulf War.

Commodore Toby Elliott RN (Retd.), who commanded the frigate HMS *Brilliant* during the Gulf War, earned my deep gratitude by offering images from his stunning photographic collection and agreeing to be interviewed.

I must thank publishers Ian Allan for putting me in touch with the family of the late R. Ransome Wallis, who was one of *London*'s medical officers during the Second World War. His son, Peter R. Wallis graciously gave permission for me to quote extensively from his father's book *Two Red Stripes* and also to reproduce photographs contained within it. It was a key contribution to the chapters on the Second World War.

The staffs of the Imperial War Museum Sound Archive and Department of Documents, the National Maritime Museum and Plymouth City Library's naval archive have been very helpful during the course of my research. The US Naval Historical Center also proved its worth again, by providing key images. Commercial sources of pictures included the ever-reliable Topham Picturepoint. Keely Storey, of the *The News* Library, Portsmouth Publishing and Printing Limited, put together a very comprehensive package of newspaper clippings concerning HMS *London* of the 1960s–1980s. It proved essential to constructing Chapter Seventeen.

Thanks are due to the copyright holders of the various collections held by the Imperial War Museum, extracts of which form an important element of this book. I am particularly grateful to the family of Commander Charles Drage, not only for allowing me to use images of HMS *London* in the First World War, taken from his superbly detailed diaries, but also for proofreading Chapters Six and Seven.

Ron Wood, who served in the next *London* on Convoy PQ18, deserves thanks for allowing me to reproduce some sketches that illustrate the grim reality of war in the Arctic.

Photographic colleagues Nigel Andrews, Tony Carney, Gary Davies, Jonathan Eastland, Harry Steele, Guy Toremans and M. Welsford provided sterling service in helping me to assemble a superb range of images for this book. Tribute is due to Stu Reed, the former Senior Public Relations Officer of Commander-in-Chief Fleet, for reminding me how exciting the trip to Russia was, via his own recollections, and for providing images and other material.

Appreciation is due to Ursula S. Carlyle, Archivist & Curator of the Mercers' Company Archives & Art Collection for finding and forwarding a fine watercolour of the cruiser *London* on the Thames in 1938, together with permission to reproduce it.

J.S. Morris very kindly provided his drawing of guided-missile destroyer *London*.

As with my book on HMS *Warspite* both Dennis Andrews and Syd Goodman have made invaluable contributions – Dennis via his superb art work and Syd by allowing me to draw on his naval archive, including original documents and photographs in the renowned Goodman Collection.

Even those who have not made it into print, via quotes about their experiences, have enriched the narrative of this book, either through providing original documents, photographs or other illuminating material. To them I give sincere thanks. If there is anyone I have forgotten to mention please accept my heartfelt apologies.

Iain Ballantyne
January 2003

ACKNOWLEDGEMENTS TO THE 2022 EDITION

A range of people helped me in a variety of ways putting together new material for this first paperback edition of *HMS London* and to whom I owe a debt of gratitude. They are (in alphabetical order): Nigel Andrews; Yorkie Cunningham; Gary Davies; Jonathan Eastland; Steve Ellis; Richard Endsor; the late Lt Col T.P. Furlonge, RMs; Peter Hore; the late Captain John Beckett Robathan RN; Ian Taylor; Ross Watton.

I'd also like to thank the public relations staff of BAE Systems Maritime – Naval Ships and NATO Maritime Command, and also the crew of the US Naval History and Heritage Command photographic archive. Of course I must express my appreciation to Pen & Sword Books for giving the green light to this new edition, not least Henry Wilson who has been very encouraging over the past two decades across various books. I'd like to thank everyone else at the publisher who made it happen too, especially Matt Jones. When it comes to those who pointed out errors in the hardback edition, and which I added to my own list of glitches for attention in this paperback, I remain appreciative of their eagle eyes. Any remaining errors of course remain my own fault.

INTRODUCTION

With Fair Winds and Following Seas

Since the last HMS *London* won the battle honour 'Kuwait 1991' and the Cold War concluded there has been no 'end of History' nor has there occurred any 'universalisation of Western liberal democracy'.[1]

On the contrary, the gears of history have continued to grind.

After a furlough in rivalry between blocs, the world has entered a new era of great power competition. It is partly expressed by a naval arms race between Russia and NATO in the Atlantic and Mediterranean and between China and the USA (plus its allies) in Indo-Asia-Pacific.

At the time this new edition was being prepared for publication war erupted in eastern Europe - via a massive Russian invasion of Ukraine - with the navies of Russia and NATO closely shadowing each other in European seas. The Kremlin also ordered its naval forces to take part in a blitz of Ukraine, firing land-attack missiles from the Black Sea at targets deep inland.

The evolution of politics and economics into new hybrids has seen the former Soviet Union transform into a capitalist, militaristic, totalitarian state run by a cabal of ex-KGB officers, with Vladimir Putin as the supreme puppet master (or 'Comrade Commander-in-Chief' as his loyal military subordinates call him).

In 1995, on the bridge wing of the Type 22 frigate HMS *London*, sailors prepare for new foes to come zooming over the horizon during a combat training exercise off southern England. *Nigel Andrews.*

Meanwhile, China – under dictator for life Xi Jinping – remains in the grip of the Chinese Communist Party, a behemoth that likewise merges pursuit of naked capitalism with Big Brother-style iron rule over the lives of its peoples.

It is engaged in a massive expansion of China's military and nurtures ambitions to depose the USA as the top superpower while achieving domination of global trade.

Rivalry on the oceans therefore continues and telling the story of how ships named *London* have fared across time illustrates that struggles between great powers have always featured key episodes involving fighting vessels and their crews. In tracing the highs and lows of the various *Londons* during a voyage spanning the centuries, this ship biography offers numerous triumphs and tragedies.

It includes adventures and misadventures in war and peace, under sail or on steam propulsion. We find out how the 98-gun *London* was there at the pivotal moment when Britain lost its American colonies in the 1780s. The same *London*'s signals were ignored by Nelson at the Battle of Copenhagen in 1801 when he defeated Napoleon's plan to strangle British trade.

In 'the long calm lee of Trafalgar'[2] we join the last wooden wall *London* bombarding Sevastopol in the 1850s, supporting the Ottomans in containing the hungry Russian bear. When the next (pre-dreadnought) *London*'s guns roared it was during bombardments of the Turks before putting ashore ANZAC troops at Gallipoli in the First World War.

The heavy cruiser *London* of the inter-war period flew the flag and upheld Pax Britannica before being remodelled for war with Hitler in the 1940s. Enduring the tragedy of the notorious PQ17 convoy, one of the blackest episodes for the Admiralty in the Second World War, she then sailed east to help end the war with Japan. In her twilight days – during the cruiser's bloodiest mission – *London* stormed up the Yangtze River in 1949 to try and rescue the frigate HMS *Amethyst* from incarceration by the Chinese communists.

As Britain struggled to maintain a global role and build a modern navy in the 1960s the men of the guided-missile destroyer *London* witnessed the withdrawal from empire. With that vessel gone from the fleet, by the late 1980s there was a new *London*, designed to safeguard Britain closer to home against the growing menace of the Soviet Union's submarines. She would enter the Baltic and the Arctic to gather intelligence and make life uncomfortable for the opposition.

That *London* achieved her crowning glory in 1990/91 as flagship of the UK naval task group that played an important role in ensuring Iraq occupiers were kicked out of Kuwait during Operation Desert Storm.

The same *London* sealed her place in history during the finale of the Cold War confrontation between the Russians and Western navies in the Barents Sea. However, with not much more than a decade in the fleet she was decommissioned due to defence cuts as British sea power suffered a precipitous decline. She was later sold to Romania in whose fleet she still sails, facing the new Russian threat.

When the first edition of this book was published, the idea of a future HMS *London* was just a dream but in 2018 it was revealed there is to be another ship of the name – one of eight City Class (Type 26) anti-submarine frigates. A new chapter at the end of this edition looks at that next *London*, laying out how she will shape up and why she is needed at the heart of a regenerated Royal Navy.

In addition to fresh imagery and some minor tweaks in the original main text, a new appendix offers a dive into fresh developments in the story of the *London* that saw service in the fleets of Oliver Cromwell and King Charles II.

The crown jewels of any ship biography such as this are the experiences of the people who sailed in them through triumph and disaster. I am delighted to enrich the human story with another new appendix, containing a few fresh stories about *London*'s veterans, especially an officer who saw combat in one of the legendary sea fights at the end of the Second World War.

*   *   *

The story of the *Londons* remains close to my heart due to my own visits to the Type 22 frigate. I first encountered that HMS *London* in the Arabian Gulf in 1990/91 and then sailed aboard her during the historic visit to Russia.

In 1995 I went aboard *London* for a day of mock-combat exercises off the Dorset coast. They were part of the process of preparing her for a deployment to the Adriatic, where I had since 1992 visited numerous naval vessels on reporting assignments off the burning Balkan shore. Those vessels included *London*'s sister vessels *Boxer* and *Brilliant,* within the range of Serb anti-ship missiles and therefore closed up at actions stations. As I write, the London of the 1980s and 1990s in which I witnessed the end of the Cold War at sea remains in the order of battle of a NATO nation. She may yet see tense patrols, and even action, with the Black Sea region plunged into a war between Russia and the Ukraine, if not yet involving NATO. We must hope that by the time this book is published it has not become an even broader conflict.

When the new HMS *London* is commissioned in the 2030s her global adventures will likely take her into harm's way too, and on missions as varied as those of her predecessors as told here. With fair winds and following seas her voyages will hopefully be equally illustrious and also lucky.*

Iain Ballantyne
August 2022

Notes

1. Francis Fukuyama made his famous analysis in an article for Summer 1989 of 'The National Interest', later expanded into his book 'The End of History and the Last Man' (The Free Press, 1992). Fukuyama has explained that what he was not suggesting the end of 'History' itself, but rather 'history understood as a single, coherent, evolutionary process, when taking into account the experience of all peoples in all times' and that even 'large and grave' events would still happen.
2. As the period of uncontested British naval supremacy between 1805 and 1914 is described by Andrew Gordon in his seminal 'The Rules of the Game' (John Murray, 1997).

---

* See the new Epilogue on p.234 for more on the naval side on the Ukraine War.

Chapter One

PRIVATEERS, ROYALISTS AND ROUNDHEADS

In the Service of King Charles

Prior to any ship called *London* serving in the Royal Navy, there were several merchant vessels carrying the name built on the Thames and crewed by sailors from local towns and villages.

In 1620 one of these early privateering *Londons* saw action against Portuguese pirates in the Arabian Gulf, fighting under the flag of the East India Company as it expanded into that area. It was entirely appropriate that a ship called *London* was one of the company's stalwarts, as the original Royal Charter signed by Queen Elizabeth I on the last day of 1600, '…gave birth to the Governor and Company of Merchants of London trading into the East Indies, commonly known as the East India Company.'[1]

The first *London* to serve with 'His Majestie's Navie Royall', as the Royal Navy was known in the early 1600s, was a 40-gun converted merchantman that was one of ten armed ships provided by the merchants of the City of London to serve in the fleet of Charles I in the late 1630s. They were required as an essential bolster to the Navy, as the King was keen to counter the growing Dutch and French naval threats.

In the summer of 1642, as civil war erupted, the *London* was among naval vessels lying at a fleet anchorage called the Downs. They were asked to declare for either King or Parliament and *London* was among eleven that immediately declared for the Parliamentarian cause. These then turned on the other five and threatened to blow them out of the water if they did not surrender. Three of them gave in immediately, but the other two refused and a tense stand-off developed that was inevitably resolved in favour of Parliament. The fleet had no real incentive for staying loyal to a monarch who had neglected it so sorely:

> *Ships rotted at their moorings for lack of attention; shortness of victuals, and pay constantly in arrears, were chronic causes of discontent among the men. In 1642, when the Civil War began, some of the sailors had received no pay for several years. Captains were reduced to selling the masts and yards of their ships to obtain money to feed and clothe their crews.*[2]

With the fleet going over to Parliament, King Charles could expect no help from overseas against the rebels. Seven years later he was beheaded and the Parliamentarians, under the iron hand of Oliver Cromwell, turned England's attention once again to securing riches overseas, where the East India Company was locked in a bitter struggle with the United Provinces' Long Distance Company. The Dutch were the giants of world trade and their merchant vessels linked the United Provinces with exotic sources of riches across the globe, from Japan and China (silk) to the West Indies (sugar cane), from the East Indies (spices) to North America and West Africa (precious metals and slaves).[3]

The Dutch killed English trading links with the Baltic and threatened to supplant them elsewhere, even in England's own backyard. They were sailing their ships into English ports to offload goods at the expense of local merchants. The temperature was further raised by Anglo-Dutch disputes over trade with Russia.

Even when Dutch republicans gained ascendancy, after the Stadtholder, William II, died in 1650, things did not improve. By then the English were pushing for unification between

Holland and England to create a Protestant superstate to present a single face against the Catholic enemies.

The Dutch viewed Cromwell's Commonwealth with some trepidation - it was a warlike monster with a fierce religious ideology they found repellant. It 'recalled the Spanish hegemony which they had heroically thrown off.'[4]

When England claimed sovereignty over the English Channel and the North Sea, it seemed the Dutch fears were well founded. Any ship that did not dip her flag in salute to English authority was attacked, and, to stamp England's authority on those waters, privateers seized nearly 200 Dutch ships in the period 1651-52. The passing of a Navigation Act '…required all imports to this country to be brought in only by English ships or those of the country of origin'.[5] This was the precursor to three naval wars that stretched across twenty-two years.

The next *London* of the Royal Navy was another converted merchant ship, blooded at Kentish Knock in September 1652, a nervous, disjointed and brutal series of scraps off the mouth of the Thames. The English fielded sixty-eight ships, under the legendary General at Sea Robert Blake, against fifty-nine Dutch warships led by Admiral Witte de With. Honours were more or less even, each side feeling the strength of the other. In this first Anglo-Dutch War, the English would hold the advantage, as their warships were generally bigger, with heavier armament, and were more numerous.[6]

London was there the following June when a massive English naval force of more than 115 ships inflicted a crushing defeat on a Dutch fleet of 104 warships at the Battle of Gabbard off the Suffolk coast. For the loss of no ships, the English took eleven Dutch vessels and destroyed nine others. A blockade of the United Provinces was enforced.

On 31 July, at Scheveningen, *London* was one of 100 English warships that clashed with a Dutch fleet of similar size off the Texel. While it resulted in the siege of the Hague being lifted, it was otherwise another crushing defeat for the Dutch. England suffered just two ships lost and 250 men killed, while the Netherlands lost eighteen and 1,600 casualties including, worst of all, Admiral Maarten Tromp, who was shot through the heart. The death of Tromp knocked all the fight out of the Dutch and Scheveningen was the last sea battle of the war. Such was English confidence in the wake of this victory that Cromwell declared himself Lord Protector of England in December 1653. But the peace agreement signed at Westminster in 1654 created no change in the situation that led to war, making it inevitable hostilities would break out again.

Oliver Cromwell's 'Lusty Ship'

The naval triumphs of the First Anglo-Dutch war were the true beginnings of the formidable Royal Navy. It was a time of great reform, which echoed what had happened ashore with the New Model Army that defeated the Royalists. Many of the Articles of War, which to this day underpin the effectiveness of the Royal Navy, were introduced. As a converted merchant vessel, the first *London* that served in Charles I's and Cromwell's navy was a bridge between the old buccaneering approach and the new military style organization. In essence, under Blake and other senior Generals at Sea, and with the full backing of Parliament and influenced by the merchants of London, who wanted more effective protection for trade, the English fleet was transformed from a collection of vessels largely in the business of privateering for the national good, into a formal fighting force. The dockyards were shaken up and reorganized, ships were supplied properly with rations and pay became a regular occurrence for the crews of the warships rather than a rarity. Naval hospitals were also established to care for sailors when they became sick.

A 64-gun, 2nd Rate of 1,104 tons, the third *London* of the English fleet was launched in the summer of 1656. The event was reported in one of the esteemed organs of Cromwell's dictatorship, the newspaper *Mercurius Politicus*. It relayed that the 'Commissioners of the Admiralty' had 'launched a lusty ship...named the *London*...'[7] It was also reported that Cromwell himself had decided the new warship's name only a fortnight before her launch. Choosing the name of England's capital city to grace her was an unusual decision, as other recent additions to the fleet had been named after famous Roundhead victories, such as Naseby. It seems his aim was to provide a tribute to a city that showed stout support for the Parliamentarian side in the Civil War. Cromwell, more than anyone, realized the value of a strong Navy as the main bulwark against invasion for an island power. This new purpose built *London* was part of a massive investment in seapower, for in 1656-57 Cromwell spent £809,000 out of a total National Revenue of just £1,050,000 on the Navy.

In August 1657, *London* first fired her guns, to mark the passing of the heroic Robert Blake. The legendary General at Sea had finally succumbed to his wounds, and his body was carried to its last resting place; *London* and other ships were on Channel guard duty.

This *London* took part in her first act of war when she helped escort English troops across to Dunkirk, which was taken as a Commonwealth possession. In 1658 Cromwell died and his son, Richard, became Lord Protector, but found it was a task beyond his powers. In 1659, he stepped down. Parliament assumed power, but believed that the nation still needed a powerful figurehead to weld everything together and so it asked the exiled Prince Charles to take the Crown.

In May 1660, the Royal Navy declared its allegiance to the new King Charles II and Samuel Pepys, Clerk of the Acts to the Navy Board, was aboard the *London* when the whole of the assembled fleet fired an exultant salute to the new ruler.

London was in a squadron of warships that sailed to the Continent to collect Charles and his court. She carried back the King's brother, James, Duke of York, who had been appointed Lord High Admiral (Commander-in-Chief). As a reward for bringing the Royal group back to England in safety, the officers and men of the escort ships were each awarded a bonus equivalent to a month's pay. It was never paid.

Pride and Joy of the Monarch
The triumphant return of the King and his brother to England trailed dismay and tragedy in its wake. While the Dutch rebuilt their maritime power to unprecedented levels and dominated world trade almost in its entirety, England paid its navy off. But, by 1665 the English had again woken up to the fact that he who has a strong Navy controls world trade and, desperate to revive the country's fortunes, were rebuilding their naval strength. The *London* was one of many neglected warships put into refit. But, in early March 1665, shortly after leaving Chatham, she blew up and three hundred people aboard her were killed instantly. Just nineteen of her crew survived and the cause of the explosion was never fully ascertained.

Within a week of the tragedy, the Lord Mayor of London and the aldermen of the city had sent a letter to King Charles offering funds to build a new *London*. As a sign of his gratitude, King Charles allowed the ship to be called *Loyall London*.

She was to be a grand man-of-war, for at Deptford they set to work building an 80-gunner with three decks. Of the £18,355 needed to build her, £16,272 was raised by subscription in the city. However, the fund raising ran out of steam. Massive corruption was exposed in the Navy's dockyards and this combined with the after effects of the Great Plague, made the citizens of London reluctant to part with their cash. A mortagage had to be arranged to find the balance and this would not be paid off until 1675.

The need for a new warship was pressing, for in the summer of 1665 war had returned and the enemy was once again the United Provinces. Cromwell's Commonwealth might have gone, but the great rivalry based on trading disputes still existed.

As *Loyall London* was being launched in early June 1666, the war was going badly for England. At the Battle of the Four-Day Fight, 1-4 June 1666 – four English warships were sunk and half a dozen captured. English casualties were severe – 4,500 dead or wounded and taken prisoner. Capitalizing on this victory, the Dutch blockaded the Thames.

But, with glorious ships like the *Loyall London* joining the fleet, surely the tide would turn? She looked magnificent – her hull and ornaments were a riot of yellow, black, blue, white and gold, but the decks and fittings out of sight were painted a grim red colour to hide the blood that would undoubtedly be spilled aboard her. But, before she could play her part in the conflict, she needed reliable weapons and *Loyall London's* entry into service was delayed because her original set of big guns shattered during firing trials. Samuel Pepys was much aggravated by this problem and desperate measures were introduced, with the battlements of forts stripped of their weapons to arm the new warship. There was also some difficulty getting together the 580 men needed to crew her. Press gangs were soon hard at work while the Army was required to send its soldiers to sea. Finally, the *Loyall London* was ready to join the fighting fleet, with Pepys going as far as calling her 'the best ship in the world'. Her first engagement would be a gruelling one – the St James's Day Fight, which was an epic battle over two days, 25 and 26 July 1666.

An English force, under Prince Rupert and the Duke of Albermarle, made up of eighty-nine warships and eighteen fireships clashed with eighty-five Dutch men-of-war and twenty fireships. The English were, in the end, triumphant, breaking the Thames blockade, sinking twenty enemy warships and killing 4,000 Dutch sailors, including four admirals, while the Royal Navy lost only one ship. The *Loyall London* was flagship of the Blue Squadron and at one stage came close to being sunk after coming under sustained fire. Disabled, she had to be towed out of the firing line. The English were so exhausted they were unable to ram home any strategic advantage and the Dutch therefore retained their grip on trade.

Both nations were weary of war and peace talks were initiated at Breda. England, still suffering the after-effects of both the plague and the Great Fire of London, together with an inevitable economic depression, swiftly paid off its expensive fleet, laying up most of the warships at Chatham. This was a foolish move, but it was unavoidable. The infrastructure that supported the Navy was in disarray. Parliament's refusal to provide further funding for a war it had come to regard as ruinous, left dockyard workers unpaid for months. The suppliers of raw materials to the dockyards also went without payment and the fleet's sailors were similarly without reward for their labours. The result was violent protest by yard workers, crews walking off their warships and suppliers refusing to provide for the fleet.

De Ruyter's raid on the Medway in 1667, with *Loyall London* (background) burning by Upnor Castle. *AJAX.*

When the Breda talks became deadlocked, the Dutch decided to take advantage of England's voluntary negligence of national defence. In June 1667, eleven months after the St James's Day Fight, Admiral De Ruyter led out seventy ships on a daring raid into the Thames. The Dutch captured Sheerness on 7 June and went up the Medway four days later. Three English men-of-war were captured and the *Loyall London*, *Royal Oak* and *Old James* were burned. As the Dutch approached, the three English warships were deliberately holed to prevent their capture as prize ships. Dutch warships,

> *...engaged Upnor Castle, their fireships tried to set fire to the three English ships sunk on the mud... they met fierce resistance and suffered casualties. They eventually succeeded in burning the upper works of the three great ships but had to abandon any thought of attacking the dockyard at Chatham or of getting at the rest of the English ships further up the river.*[8]

In the meantime defences up the Thames were being stiffened to prevent the Dutch from attacking London itself. Luckily they were deterred and the capital was saved. The day on which *Loyall London* was burned has been described by one historian as '...probably the blackest day in English naval history...'[9]

In the daring adventure up the Thames and Medway, the Dutch lost just two ships and sailed away with the greatest prize of all – the *Royal Charles* (previously the *Naseby*), the pride and joy of the English fleet. At the end of July 1667 the Treaty of Breda was signed, bringing the war to an end, with the English utterly exhausted and dispirited. The Dutch were allowed to trade freely with English ports and pass through the North Sea and Channel without being harassed unjustly. The English retained one great prize they had seized during the war – possession of New Amsterdam and the New Netherlands in North America.

Meanwhile, King Charles, who had been severely dismayed by the *Loyall London*'s cruel fate, paid for her to be raised and restored. This was possible because most of the ship's structure had been protected from fire below water. The job took three years and a further £20,000 with an extra sixteen guns added to her armament. When it became clear that no

5

funds were going to be forthcoming from the people of London to help in the project, the King decided she should henceforth be called *London* – the 'Loyall' being removed as a sign of his anger. It is not surprising the good folk of London were reluctant to fund the vessel's revival, as they were still paying off the original mortgage on her construction. On top of that, the unbroken Dutch dominance of maritime trading meant that money continued to be scarce.

The struggle to prevent the Dutch strangling English prosperity would lead to war again. In addition to trading disputes, King Charles was very angry about a painting of *Loyall London* burning on the Medway, that had been hung in the Dordrecht Council Chamber. And, if the depiction of his favourite ship ablaze was not enough, the King was further incensed by an inset painting showing him in a state of partial undress with a half-naked woman on each knee.

The rebuilt *London* joined the fleet in the summer of 1670, just as the King was becoming annoyed by another insult to the old *Loyall London* – a medal minted by the Dutch to celebrate the destruction of the warship. When Dutch vessels failed to salute properly one of his Royal Yachts sailing just off the coast of the United Provinces, King Charles decided to provoke war. Hostilities opened with an English attack on a Dutch merchant convoy off the Isle of Wight. England intended to wrest control of world trade from the Dutch and had agreed to join forces with France, the latter intending to send an army to occupy the United Provinces.

The naval force supporting the invading troops was assembled in early May 1672, but then wasted too much time getting ready. The Dutch launched a pre-emptive strike, falling on the Anglo-French fleet while it was at anchor in Sole Bay, off the Suffolk coast. It had been assumed that the Dutch were still in their home ports, so it was a shock to see them coming over the horizon. *London* was at the forefront of the action, on 28 May, as flagship of Vice Admiral Sir Edward Spragge and as flag vessel to the Duke of York when his original, the *Royal Prince*, was badly damaged. At one point *London* '…was engaged by three Dutch ships and survived only after the loss of her mainmast and 200 crew'.[10]

The apocalyptic sound of the cannons firing carried to Cambridge, where Sir Isaac Newton cocked an ear and declared the English to be winning, his assumption based on the fact that the firing was gradually getting further away. In reality, the Anglo-French force was badly coordinated, having never recovered its composure and suffered some considerable

The confused opening moments of Sole Bay. HMS *London* is fourth from the right.
Illustration by Dennis Andrews.

damage. Led by *London*, still carrying the Duke of York, it pursued the Dutch, but thick fog prevented hostilities from being renewed. The wily De Ruyter had achieved a strategic victory:

> De Ruyter had disabled the English fleet for about a month, had wrested command of the Channel and thwarted an invasion of the Netherlands.[11]

The English had suffered the loss of four ships, with 2,500 men killed while the Dutch lost three vessels. The French went ahead with their land invasion and came close to success, but were thwarted when the Dutch flooded the flatlands. The shock of nearly being overwhelmed by the French was so severe for the United Provinces that the ruling republicans were deposed and the Prince of Orange was declared Stadtholder (national leader).

Joining forces with a French squadron again, the Royal Navy planned another foray. The Battle of Schooneveld, on 28 May 1673, was a nine hour contest in which the French lost two ships and the Dutch one. The defenders retained the strategic advantage, having prevented troops from being landed in the Scheldt estuary.

Another inconclusive clash occurred on 4 June, which gave the Dutch the benefit of driving the English back home to get supplies and make good damage. As most English ships, including *London*, had been sent to sea lacking in almost every respect, from a shortage of sailors to a scarcity of spare topmasts, there was lots of work to be done. Sickness and desertions soon whittled the fleet's manpower down to a pitiful state and there was no beer, bread or meat worth eating aboard the ships. In the meantime the Dutch fleet was at sea in force.

Somehow, in mid-July another Anglo-French fleet, of nearly 150 vessels, carrying several thousand troops, sailed forth to attempt once again to subjugate the United Provinces. With this combined fleet hovering off their coast, the Dutch decided to seek battle and destroy it before it could land its troops. And so Texel, the final naval battle of the third Anglo-Dutch War, started on 11 August, turning out to be another engagement in which tactical honours were even. But the Dutch retained the strategic advantage through preventing the landing of troops.

The Glorious Revolution
In England there was increasing unease about the alliance with France, particularly as the French naval squadrons had contributed so little to encounters in which they had been involved. The main purpose of the French presence seemed to be to observe Protestant nations tearing each other apart.

Weary of war, the King decided to give Dutch peace proposals a positive response. The Treaty of Westminster of early 1674 was the result. Among other things, the United Provinces agreed to English sovereignty over waters off the south and east coasts of England. They also returned New York (as New Amsterdam was now known) which they had taken back temporarily.

In the fourteen years following the Treaty of Westminster, corruption again became endemic in England's dockyards. For example, six years after Texel, holes in *London*'s hull caused by Dutch cannon fire had still not been repaired. The dockyard officials received money to carry out the work, but it went into their pockets. There were also disturbing undercurrents affecting the monarchy. Charles II had signed a secret treaty with Louis XIV of France in 1670 that provided funding, independent of Parliament, for the war against Holland, in return for an increase in Catholic influence in England at a future date. In the closing months of his life King Charles at least started to reform the administration of the Navy and this work was carried by his brother when he became James II.

But, in 1669 the Duke of York had become a Catholic and, on succeeding his brother in 1685, King James II began a process of bringing his faith to a more prominent position. With the birth of a son, it looked like England might come under the rule of a Catholic dynasty and this prospect stirred up violent objections, particularly in the Royal Navy, which was noted for its anti-Catholicism.

In June 1688, a delegation led by an admiral was sent to seek an audience with Holland's ruler to ask for immediate help. King William was married to the British monarch's eldest daughter, Mary, and this, together with his royal English blood (his mother was a daughter of Charles I), gave him a powerful claim to the throne. William began assembling an army and powerful fleet to achieve the overthrow of his father-in-law.

When France became embroiled in a campaign of conquest in Germany in November 1688, William unleashed his forces. The English fleet, most of it unwilling to fight for King James, did little to hold back William's 450 invasion ships, when twenty thousand Dutch, English and Scottish soldiers landed in Devon to march on London and seize power. The Royal Navy declared its allegiance to William and Mary in mid-December and they were made King and Queen in February 1689. The following month the deposed King James was landed in Ireland by French warships, together with an army provided by King Louis, to initiate a campaign to regain the throne. It failed and James returned to France to try and raise another army. England declared war on France in May and a combined Anglo-Dutch fleet faced the prospect of an immense French invasion force. William decided the best form of defence would be attack and so began drawing up plans to invade France. In the meantime the Anglo-Dutch fleet, including *London*, was at sea in strength to keep the French at bay. After some inconclusive encounters, the decisive moment came with the Battle of Barfleur-La Hogue on successive days in May 1692.

The French appeared to be concentrating their invasion troops and transportation ships at Vaast-La-Hogue in the Contentin peninsula and French warships destined to escort the troop carrying vessels across to England were sighted off Cap Barfleur, with battle joined on 19 May.

In fact, the French had only half their battle fleet at sea. Such was Louis XIV's impatience to get the invasion underway, that he had ordered it to sea before it was ready. The concentrated Anglo-Dutch fleet was therefore given the instant advantage of outnumbering the enemy nearly two to one. The encounter was extremely bloody and in the centre of the battle, there were sixteen French ships pitted against twenty-seven English (1,150 guns against 2,000).[12]

The *London*, flagship of Sir Cloudesley Shovell, was in the middle of the maelstrom but she was well prepared for action:

> *In her magazines the* London *carried ten tons of powder and twenty-five tons of shot… Hand grenades, fused ready, were on the quarter-deck, in case it came to boarding…*[13]

But, the two French men-of-war opposite *London* appeared to have the advantage:

> *…the 84-gun* St Philippe *and the 90-gun* Admirable *opposed the 96-gun* London. *With the sea barely ruffled by the dying breezes, all three tiers of gun port lids were up; the great guns beneath erupted in smoke and darts of flame; round shot flew in pairs between the floating fortresses, ploughing through the massy oak sides, launching showers of splinters across the decks, smashing flesh and bone, gouging masts, snapping rigging, tearing holes in canvas.*[14]

The *London* swiftly reduced the *Admirable* to a blazing wreck, while the *St Philippe* was grievously injured. The French tried to escape north, but the Anglo-Dutch fleet gave chase. Some of the French vessels tried to evade their pursuers in Vaast-La-Hogue, but the English followed them in and inflicted more carnage. From the nearby shore, King James looked on with bitter admiration as the fleet he had once commanded destroyed any hopes he might have of regaining the throne of England.

The score sheet at the end of this momentous clash was fifteen French vessels lost for no ship losses on the Anglo-Dutch side. In the aftermath, the French battle fleet declined rapidly, with King Louis diverting all his resources into a huge army that made France the most dominant power on the Continent. Meanwhile, the Royal Navy expanded in a manner commensurate with English trade, which now thrived as never before.

By October 1694, *London* was the Nore Guardship, the administrative flagship for the naval vessels on the Thames and in the Medway and the oldest man-of-war afloat in the British fleet. In October 1695 the *London* was paid off and taken into dry dock to make her fit for sea service again and she finished the war taking part in blockades of ports in the Contentin Peninsula.

In 1701, the Admiralty ordered that the *London* should be reconstructed at Chatham. She was therefore dismantled and carefully rebuilt using as many of her old timbers as was realistically possible. The process was not pursued with any great vigour and *London* was not floated until the spring of 1706. She became the flagship of Commander-in-Chief Nore and was charged with leading the rest of the ships at anchor in celebrating the Union of England and Scotland on 1 May 1707. The *London*'s log remarked:

> *Thursday, May 1st, 1707. – At 1 p.m. fired 21 guns being ye Day appointed for ye Cellebration of ye Union between England and Scotland, we having all our fflags and Colours and pendants fflying.*[15]

The *London* received one final rebuild in 1722, her dimensions expanding again, for in her long life she '…grew from her original 1134 tons to an eventual 1711 tons…'[16]

By 1746 the fourth *London* had become a floating chapel, and the following year was sent for breaking up at Chatham. She had recorded eighty-one years' distinguished service.

Notes

1. Antony Wild, *The East India Company*.
2. A. Cecil Hampshire, *A Short History of The Royal Navy*.
3. Roger Hainsworth and Christine Churches, *The Anglo-Dutch Naval Wars 1652-1674*.
4. ibid.
5. A. Cecil Hampshire, op. cit.
6. David A. Thomas, *Battles and Honours of the Royal Navy*.
7. Edward Fraser, *The Londons of the British Fleet*.
8. Frank Kitson, *Prince Rupert Admiral and General-at-Sea*.
9. Peter Padfield, *Maritime Supremacy and the Opening of the Western Mind*.
10. Taken from an article entitled *The Ships That Chatham Built*, published in *Periscope*, the newspaper of Chatham Royal Dockyard, April 1980.
11. David A. Thomas, op. cit.
12. Peter Padfield, op. cit.
13. Edward Fraser, *The Londons of the British Fleet*.
14. Peter Padfield, op. cit.
15. Edward Fraser, op. cit.
16. *The Ships That Chatham Built*, *Periscope*, April 1980.

Chapter Two

THE SHIP THAT LOST AMERICA

The *Victory*'s Contemporary

An 80-ton ship armed with only eight guns, the fifth *London* was a converted merchant vessel purchased by the Admiralty and in April 1758 participated in an expedition to capture the French colony of Senegal in West Africa. She was wrecked on a sand bar in the mouth of the River Senegal as the British squadron moved in to attack the stronghold of Fort Louis. Luckily, loss of such a minor vessel could not prevent the attack from succeeding.

A 98-gun 2nd Rate, the sixth HMS *London* was to have a life spanning the highs and lows of British naval power. Admiralty authority for her construction was given in December 1758, the year of Nelson's birth and she was built at Chatham alongside the *Victory*. At 1,894 tons, she was just under 300 tons lighter than *Victory* and woods used in her construction included fir, elm and oak. Launched at the end of May 1766, she bore the arms of the City of London carved under her figurehead. The Seven Years War had ended just three years earlier, a conflict that pitted Britain against France and Spain in a war that was primarily about overseas possessions, mainly in North America, India and the Mediterranean. For Britain the war consolidated colonies in North America and the West Indies, while furthering the influence of the East India Company and reinforcing a presence in the Mediterranean. The Royal Navy had been masterful, ruling the waves for King George II. His grandson, who came to the throne in 1760 as George III, pursued the war his father had started.

But, in the aftermath of victory in the Seven Years War, as was customary, the Royal Navy was run down. The war had been very expensive and drastic economies were needed to keep the state solvent. In fact the Royal Navy was cut back to such an extent that it could no longer be confident of countering the Franco-Spanish threat.

A line engraving of HMS *London* off Rame Head, made in 1781. *US Naval History and Heritage Command.*

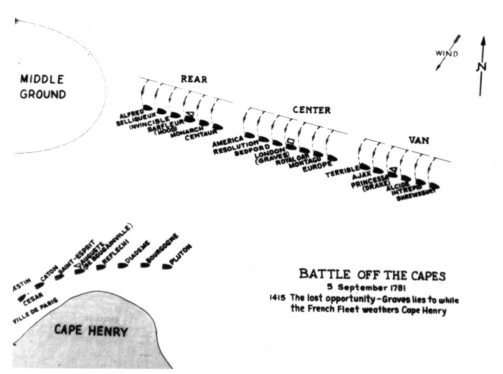

MIDDLE GROUND

REAR

CENTER

VAN

WIND

N

ALFRED
BELLIQUEUX
INVINCIBLE
BARFLEUR
(HOOD)
MONARCH
CENTAUR
AMERICA
RESOLUTION
BEDFORD
LONDON
(GRAVES)
ROYAL OAK
MONTAGU
EUROPE
TERRIBLE
AJAX
PRINCESSA
(DRAKE)
ALCIDE
INTREPID
SHREWSBURY

ASTIN
CATON
SAINT-ESPRIT
AUGUSTE
(DE BOUGAINVILLE)
REFLECHI
DIADEME
BOURGOGNE
PLUTON
CESAR
VILLE DE PARIS

CAPE HENRY

BATTLE OFF THE CAPES
5 September 1781
1415 The lost opportunity – Graves lies to while
the French Fleet weathers Cape Henry

The Battle of the Chesapeake Capes, 5 September 1781, the clash that HMS *London*'s honours board does not celebrate. *US Naval History and Heritage Command/NH 73449.*

An opportunity for avenging defeat in the Seven Years War was provided for France and Spain by King George's own subjects in America whom they helped in their fight for independence. In the spring of 1778 Britain declared war on France over its policy of supporting the American rebels, but HMS *London* was not ready for the first encounters between the French and British navies. She spent many weeks in dry dock at Chatham having rotten timbers replaced and would not leave dockyard hands until November 1778. Indeed many Royal Navy warships were in a state of disrepair and the dockyards had in recent years become riddled with inefficiency and corruption.

In August 1779, *London* and other ships of the Royal Navy scoured the seas searching for a Franco-Spanish Combined Fleet. The ninety vessels of this powerful force were ultimately discovered anchored just off Plymouth. With the British 'Western Squadron' of sixty ships more than 100 miles away to the west, it seemed the Devon naval bastion was doomed, as its defences were in a lamentable state. The Franco-Spanish objective was to seize and hold Plymouth '...as hostage for the return of Gibraltar...'[1] which had been captured by the British during the Seven Years War.

The commanders of the Combined Fleet were not aware of the poor state of the English naval port's defences and disease and starvation was taking its toll of their poorly provisioned ships. Leaving without mounting any kind of landing, the Combined Fleet headed west, in order to seek battle with the Royal Navy. The two fleets often came within sight of each other, but neither one could manage to close before the other disappeared again. This shadow boxing went on into September when the French went their way and the Spanish theirs.

The war at sea went into hibernation at the end of November and, with the onset of spring, HMS *London* found herself assigned to fly the flag of Rear Admiral Thomas Graves on the North American station.

In May 1780, *London* set out in company with five other men-of-war to make the transatlantic crossing. Their journey was given urgency by the fact that they were chasing seven French men-of-war and four frigates carrying 6,000 troops destined to reinforce anti-British forces in North America. But the French vessels managed to evade their pursuers, landing soldiers at Rhode Island, which they swiftly captured along with the harbour at Newport. *London* and the other ships arrived in July and helped Royal Navy vessels already on station to mount a blockade to keep the French bottled up.

At that time British land forces were split into two armies – one based around New York, under General Henry Clinton, and the other in Virginia, under General Charles Cornwallis. Sea was the vital link, as rebels held the territory between them. Luckily the Royal Navy seemed to have the situation under control, so the two armies would continue to be supplied and reinforced from the sea, while any daring moves by the French to bolster their own forces could, hopefully, be easily destroyed.

The naval battles that decided the issue, and therefore the whole war, took place in Chesapeake Bay. The first encounter, in March 1781, was the one that gave *London* her Chesapeake battle honour, despite being a less remarkable affair than the second. Seven French ships were carrying two thousand troops they intended landing to reinforce the rebels fighting Cornwallis. Eight British warships, including *London*, sought to drive off the French and they succeeded in preventing a landing, but at considerable cost – three of the Royal Navy vessels were dismasted while the *London* never really got into the action. The French retreated to Newport and the British would have given chase but for the extensive damage they had suffered.

To seize the initiative and close the trap on the British army in Virginia, which had retreated to Yorktown, the French decided to send twenty-four ships of the line, under the command of the Comte de Grasse, north from the Caribbean. With eight ships of the line already there, the French would be able to establish overwhelming superiority as the British could only gather together nineteen ships in North American waters. This did not prevent the British from sailing south from New York to attack French warships that had landed troops and were blockading Yorktown. The British had the advantage of surprise on 5 September, but Rear Admiral Graves, flying his flag in HMS *London*, failed to press his attack home with enough urgency.

Graves chose to hold off while he formed a perfect line of battle as laid down in the Admiralty's *Fighting Instructions*. The two converging fleets made a V-shape, and the French were able to concentrate more firepower at the head of their line. The *London* herself, tenth in line, had three of her guns knocked from their mountings and suffered more than twenty casualties. The French did not want to engage the enemy too closely, as their main purpose was to maintain lines of supply to the rebel army ashore. They withdrew, leaving the British to lick their wounds and ponder the missed opportunities.

When Rear Admiral Graves held a meeting of his commanding officers aboard *London*, the full extent of the damage to several of his vessels was revealed. One, HMS *Terrible*, had been suffering leaks prior to the battle and was now in a perilous state. She ended up being abandoned and burned.

The Rear Admiral now sat down in his cabin aboard *London* and wrote his after action report, which concluded by saying:

> *In the present state of the fleet, and being five sail of the line less in number than the enemy, and they having advanced very much in the wind upon us during the day, I determined to tack after eight, to prevent being drawn too far from the Chesapeake, and to stand to the Northward...*[2]

Graves should now have taken his fleet into Chesapeake Bay, but instead stayed outside, leaving the way open for the French. In the meantime de Grasse took his ships to meet the French squadron from Rhode Island, which was carrying troop reinforcements and siege guns. They were able to sail back into Chesapeake Bay and make an unhindered landing. De Grasse now had thirty-five ships of the line, while Graves had only eighteen. Rear Admiral Graves held a council of war aboard *London* where it was decided the best course of action would be to return to New York for repairs and await the arrival of more ships from the Caribbean and England. However, such was the Franco-Spanish threat in home waters and the West Indies that only three ships of the line could be spared and the inefficient dockyard at New York took a perilously long time to conduct the repairs. The siege of Yorktown tightened by the day, with the new guns pounding British defences to pieces. To remedy the situation, it was proposed the British fleet in North America should sail south carrying 5,000 troops, break the naval blockade and raise the siege. By the time the ships were ready to sail in mid-October it was too late. On the day *London* and the others set sail on their rescue mission, General Cornwallis was signing the formal surrender of his army at Yorktown. When the British ships arrived off Chesapeake they could tell by the eerie silence that the battle was lost. One of Rear Admiral Graves' subordinate commanders at Chesapeake, Admiral Samuel Hood, complained bitterly about the lost opportunities. But even so, neither he nor Graves would have realized at the time how the botched battle would lose Britain the American colonies. To the Admirals it had been just another engagement in a long war. But, it was the second time a British army had surrendered to the rebels – the first being at Saratoga in 1777 – and it was the last straw for a nation weary of the financial burden of the war and political turmoil.

Meanwhile, a new threat to Britain's rich Caribbean colonies was emerging. The Comte de Grasse was ordered to bring his fleet south again to the West Indies, where he was to join up with nineteen major warships dispatched from France carrying 8,000 troops and siege guns. Spain contributed twenty ships of the line and 5,000 soldiers, the aim of this combined fleet being the capture of Jamaica. When Sir George Rodney took a British fleet out to seek the enemy off Dominica, he managed to find the main French force on advantageous terms. In the subsequent Battle of The Saints, on 12 April 1782, the British fielded thirty-six ships of the line against thirty-two French. It was a resounding Royal Navy victory due to superior British gunnery, sailing and discipline. The French lost five ships and suffered more than 2,000 casualties, while the British, who sustained 1,000 casualties, lost not a single ship. The Comte de Grasse was captured along with his flagship, *Ville de Paris*. The Franco-Spanish invasion force remained poised but its teeth had been drawn and the courage of its commanders ebbed away.

When Rodney arrived at Port Royal in the wake of his victory, he was greeted by HMS *London* fulfilling her duties as Jamaica station flagship by firing a seventeen-gun salute. The *London* had arrived at Port Royal in December 1781 still carrying Rear Admiral Graves. His *Log Book of Proceedings of His Majesty's Ship London* records *London*'s actions on Rodney's return:

A ghostly figure walks the gundeck of HMS *Victory*, which HMS *London* 98-guns was built alongside at Chatham. *Victory* was a 104-gun 1st Rate and *London* a 2nd Rate, but this image gives a good idea of her own gun decks. *Jonathan Eastland/AJAX.*

Pm at 6 saw Admrl Rodney with 21 sail in compy, to windward. Hove short at ¹/₂ past got under way...at 5 saw Admrl Rodney wiy 16 sail in co., at 7 saluted him with 17 guns. Hoisted out the Barge. ¹/₂ past Admrl Graves went on board. Admrl Rodney at 11 made sail.[3]

The *London* put to sea with other ships of the fleet shortly after Rodney's return, to watch out for any revival of the enemy's invasion plans, but found no action that spring and summer.

In October 1782, the *London* was patrolling off San Domingo, in company with the 74-gun HMS *Torbay* and the sloop HMS *Badger*. On the morning of 17 October they gave chase to two French warships – the 74-gun *Sybille* and the frigate *Scipion* – with the *London* drawing ahead of her companions to find herself in a running fight on her own. The French kept up a good rate of fire with their stern chase guns, while, due to clever seamanship, the *London* was able to remain in close contact, yawing to fire the occasional broadside. The fight lasted into the evening and, at one point, the *London* and *Scipion* collided and became entangled. On shaking herself loose, the *Scipion* fell behind *London*, managing to pour a devastating fire into the British ship's stern. In the meantime the *Sybille* had escaped while the *Torbay* caught up, guided by the flashes of gunfire ripping the night apart. On entering the fray, she narrowly missed colliding with the *London*. The *Scipion*, having run out of powder, took advantage of the confusion and slipped away. The next morning revealed the *Scipion* to her pursuers once more and this time *Torbay* led the chase, with *London* close behind. The desperate French ship tried to find refuge in the bay of a nearby island but struck rocks and sank. Although a small-scale action, which received no battle honour, the fight with the *Scipion* cost the *London* 100 casualties and some serious damage. She was, however, able to reach Port Royal in safety and, after repairs, was able to resume patrolling duties. HMS *London* saw no further action in the Caribbean and, following the Peace of Versailles that saw America securing its independence, she was called home, arriving back in British waters at the end of July 1784 to be decommissioned at Chatham.

Notes

1. Peter Padfield, op. cit.
2. Papers of Admiral 1st Baron Graves, National Maritime Museum.
3. ibid.

Chapter Three

MUTINY AT SPITHEAD

At War With Revolutionary France

In 1795 Britain found herself at war with Revolutionary France and HMS *London* was part of the Channel Fleet. That June, under the command of Viscount Bridport, the fleet chased down and defeated a French squadron off Quiberon Bay at Île de Croix. Three of the French ships struck their colours, among them *L'Alexandre*, which had been brought to heel by the guns of *London*.

London was flagship of Vice Admiral John Colpoys at Île de Croix and carried him again in December 1796, as he commanded fifteen ships of the line attempting to keep a tight watch on the enemy fleet at Brest in atrocious weather. Having been blown many miles off station, the British ships were unable to prevent the French fleet from leaving Brest, carrying thousands of troops that were to be landed in Ireland to start a rebellion. But the same bad weather that drove HMS *London* and the other Royal Navy vessels far to the west scattered the French invasion ships.

The following spring *London* gained notoriety as the only ship involved in the Spithead Mutiny upon which fatalities occurred. The mutiny started in mid-April and, at that time,

Admiral Sir John Colpoys, who was faced with mutiny aboard HMS *London*. *US Naval History and Heritage Command/NH 65903.*

did not infect the *London*. She was a fairly well-disciplined ship and, in common with the rest of the ships had an order book to supervise the conduct of her sailors. Entitled *Additional Rules and orders for the better government of His Majesty's Ship London*, it listed thirty-six rules, which included:

> *2nd The Lieutenant of the watch from six to eight to see that all the lights are out. The Hammocks hung up and everything quiet below and report the same to me as the commanding officer.*
>
> *4th The Lieutenant at arms to have the Top men and Boats crews particularly trained to the use of small arms.*
>
> *7th Cursing and swearing to be discouraged and checked at all times, but any use of that scandalous swear word <u>Buggar</u> will certainly be punished.*

The 9th additional rule advised most sternly that 'money, clothes or other articles' found on the upper deck should be brought to the quarterdeck, and anyone who did not could be accused of theft.

The instructions also included rules about what to do 'On Alarm of Fire' including: 'The Carpenter and his crew to be ready on the lower deck with their tools.'[1] But rather than harsh discipline, the main grievances that sparked the disturbance at Spithead were disgraceful pay and rotten food. While the Army's wages had been raised to keep pace with inflation since the beginning of the new war with France, there was nothing offered to the sailors, who had not seen their income increased since the reign of Charles II. Inflation made their pay worthless and the fact that merchant sailors were paid much more aggravated the situation.

The Royal Navy's seamen were fed salted beef decades old, their cheese was so hard it was used to carve ornamental boxes, shore leave was rare and sick matelots were being neglected. Many of the new recruits to the Navy's ranks were neither mild in temper nor dumb enough to tolerate such conditions for long. The 1795 Quota Acts had required communities across Britain to contribute men for naval service and the press gangs were also hard at work. Such was the pressure for quotas to be met that towns cleared out their prisons and sent hardened criminals to the Navy, along with other misfits. Some of the new 'recruits' could even read, and were knowledgeable about the rights of man that had been preached during the struggles of the American War of Independence and the French Revolution.

In March 1797, the men of the Channel Fleet drew up petitions calling for better conditions of service and smuggled them to the retired Admiral Lord Howe. Although a strict disciplinarian, he was popular among the men and he took their petitions to the Admiralty, which did nothing. The sailors were dismayed, for they believed Lord Howe had failed to communicate their grievances at all. They drew up more petitions to be sent directly to Parliament and the Admiralty. On catching wind of this, the Admiralty ordered the Channel Fleet to sail. However, when Lord Bridport gave the order to weigh anchor on 17 April, nothing happened. The mutineers stated that they '...would maintain the ships in perfect order and discipline...but they would not proceed to sea in accordance with the Commander-in-Chief's orders.'[2] Over the next fortnight they made their grievances plainly known in meetings with senior admirals. They demanded an Act of Parliament to institute better conditions and wanted the King's pardon to save them from punishment. They believed these terms had been agreed properly, but no action was forthcoming, so the mutiny flared up again. This time HMS *London* became involved.

On 7 May ringleaders from the other ships in the fleet rowed across to ask her crew to join them. Vice Admiral Colpoys, by now second in command of the Channel Fleet, had earlier observed the red flag of rebellion being run up on the other warships and, as the rowing boats carrying the mutineers approached his flagship, he decided to make a direct appeal to the *London*'s crew not to mutiny. They appeared to respond as many of the sailors went meekly below decks.

Colpoys' next step was to muster the marines and officers on *London*'s upper deck and prepare to repel the mutineers, but, as the marines went to guard the ladders, some of *London*'s own sailors stood rebellious on the ship's forecastle, refusing to go below. On coming alongside, the mutiny ringleaders shouted up to the rebels on *London*'s upper deck and appealed to other sailors looking out from gun ports, urging them to take control of the ship. The rebels on the forecastle started to turn one of *London*'s carronades around so it could fire on the officers and marines gathered aft. The First Lieutenant ordered them to stop immediately or he would fire. Some of the sailors responded to this threat by leaping back from the cannon, but others carried on turning it. One of them told the First Lieutenant to fire his pistol if he dared. He did, and the sailor dropped to the deck with a fatal wound.

Now the entire lower deck of the *London* rebelled. The sailors on the forecastle stormed aft, overpowering the officers and those few marines remaining loyal. Marines guarding the ladders threw down their muskets and the mutiny ringleaders came aboard. As sailors stormed up from below decks, punches were thrown and shots fired. A marine officer, a seaman and a midshipman were wounded and at least one other ordinary rating was killed. The First Lieutenant was seized, a noose put around his neck and the rope flung over a yardarm. The mutineers intended to reap some revenge for their dead shipmate. But one of the sailors who had come aboard from the boats shouted it would be wrong to hang the First Lieutenant, as he had served with him and knew him to be brave and honest. Admiral Colpoys called for reason, telling the mutineers they might as well hang him, as the First Lieutenant had been acting on his orders. He added that the officers had been obliged to suppress the rebellion by Admiralty standing orders. No one wanted to go as far as stringing up the Admiral and the wiser heads among the sailors held sway. The First Lieutenant's head was taken out of the noose and all the officers were confined to their cabins while the mutineers discussed what to do with them.

When word spread around the fleet that there had been fatalities among the ordinary sailors in *London*, hotheads began demanding 'blood for blood!' But the crew of the *London* refused the call and declared that no harm would come to their officers, especially the First Lieutenant. They believed he had only acted properly in order to prevent greater bloodshed.

Every one of the mutineers understood the unwritten laws of dissent in the Georgian fleet – mutinies that avoided violence and in which naval regulations were by and large observed, more often than not resulted in their Lords of the Admiralty eventually taking action to right the perceived wrongs. Mutinies in which violence was used, particularly against officers, and any insubordination when the enemy was close at hand, were treated with the utmost severity.

The midshipman, the marine officer and the wounded sailor were sent ashore for treatment in a naval hospital, while the Admiral and other officers were kept confined in their cabins for three days before being put off the ship.

In April and May 1797 Britain was wide open, with the Channel Fleet obviously incapable of defending its shores. The news of the violence aboard HMS *London* seemed to be a

Sailors of the Channel Fleet man the yards of a warship during the mutiny at Spithead in 1797.
AJAX Vintage Picture Library Collection.

precursor to the mass bloodletting that would bring Britain to its knees. But a Seamans' Bill was passed in Parliament that met all the sailors' demands for better pay, food and care for sick sailors. A Royal Pardon was passed by the King, who squarely put the blame for the mutiny on senior officers:

> It was a famous victory for ordinary men who combined irresistibly to get their rights. Apart from the trouble in the London, there was no bloodshed. No Spithead mutineer was court-martialled, imprisoned, flogged, sentenced to death or hanged.[3]

When a Coroner's Court was convened to consider the circumstances in which the sailor was killed, the First Lieutenant was declared to have committed 'justifiable homicide'. This verdict was greeted with cheers when the officer returned to the London and his popularity was higher than ever among the sailors. His bravery in shooting a mutineer had unintentionally been the key to the mutiny's success. Vice Admiral Colpoys never held another sea going command and was villified in a contemporary ballad:

> The murdering Colpoys, Vice Admiral of the blue,
> Gave order to fire on the London ship's crew...[4]

Following the mutiny, London was ordered to join the Mediterranean Fleet, which was notorious for its harsh discipline, but also recognized as the premier fighting force of the Royal Navy, under the command of the Earl of St Vincent. When the London set sail under a new captain she was still a sullen ship, but, on nearing the waters off Cadiz, which the fleet was blockading, her sailors became more compliant.

The novelist Jane Austen had a brother named Francis serving as First Lieutenant of HMS London in 1798, having replaced the hero of the Spithead Mutiny. He wrote frequently to Jane, his letters going via Lisbon in Portugal after being collected by boat. Francis often relayed details of the harsh discipline imposed by the Commander-in-Chief. He painted a vivid description of one sailor being 'flogged around the fleet' after receiving punishment aboard the London:

> The small procession then continued on its way, the prisoner to receive his quota of lashes alongside each ship of the Fleet. In the London according to the regulations, the Articles of War were read to the crew and they returned to their routine of daily tasks, washing down 'the middle and lower decks'.[5]

The London's time with the Mediterranean Fleet was bereft of action though British warships came close to bringing both the Spanish and the French fleets to action more than once, sometimes giving chase for months in vain. As the century entered its final days, HMS London was called home to serve in the Channel Fleet again and not far over the horizon lay a battle where part of the Nelson legend was forged.

Notes

1. *Orderbook of Additional Rules and orders for the better government of HMS* London, National Maritime Museum.
2. Richard Woodman, *The Sea Warriors*.
3. John Winton, *An Illustrated History of the Royal Navy*.
4. C. Northcote Parkinson, *Britannia Rules*.
5. Brian Southam, *Jane Austen and the Royal Navy*.

Chapter Four

NELSON TURNS A BLIND EYE

A Ball for the Admiral's Wife

Embroiled in yet another potentially draining war against the French, the British were desperate to ensure their enemy was isolated and weakened. But Russia, Denmark and Sweden were determined to carry on trading with France. Under the leadership of Russia's Tsar Paul, they had decided to revive their League of Armed Neutrality, which would robustly resist any efforts by the British to prevent them trading with France. Flushed with confidence from the victories of his armies, the French dictator Napoleon Bonaparte reinforced the Baltic powers' unfavourable opinion of the British, by ordering his warships not to stop and search their merchant ships for contraband, as was usually the custom with vessels belonging to neutral powers. The British refused to return the gesture. In response, the Russians went as far as impounding all British merchant vessels in their ports. Alienation from the Baltic states was a serious blow. Timber needed for the construction of warships came from the Baltic, along with hemp for naval rope making and flax for sails. With the Channel Fleet constantly at sea, running a close blockade off the main French naval port of Brest, there was an urgent need '...for constant supplies...' of fir, hemp and flax. 'Stop these items and the blockade would come to an end.'[1] That would leave the way open for an invasion of Britain.

By early March 1801 a squadron was being assembled by the Royal Navy to destroy the League of Armed Neutrality. Britain's fleet was confident it would succeed, as it was toughened by decades of war, with many seasoned ships' captains and admirals. However, not every British admiral was a fire-eater like Vice Admiral Sir Horatio Nelson, the hero of the Battle of the Nile who was still worshipped throughout the land three years after that epic victory. Unfortunately he was only second in command of the Baltic Squadron and no one could be further from the Nelsonian mould than the man who had been appointed its commander. Admiral Sir Hyde Parker, was '...kindly, fussy, stout, rubicund and very rich, he looked more like a yeoman farmer than a naval officer...'[2]

An engraving of the view over the sea to Copenhagen around 1801.
US Naval History and Heritage Command/NH 66002.

Organizing a glittering ball for his new teenage wife was more absorbing for the sixty-four-year-old Admiral than final preparations for waging war in the Baltic. His accommodation, until the flagship HMS *London* arrived from Portsmouth, was a hotel called The Wrestlers Arms. But, even when the flagship dropped anchor off Yarmouth on 9 March, the Admiral still failed to leave the warm embrace of his plump young wife.

Aboard HMS *St George*, Nelson grew increasingly agitated. He is said to have remarked bitterly: 'How nice it must be laying abed with a young wife to a damned cold raw wind.'[3]

Nelson realized that, with every passing day, the waters of the Baltic warmed and ice, blocking naval ports beyond Copenhagen, would soon melt, allowing the Danes, Swedes and Russians to combine forces. The Royal Navy had to roll them up swiftly one at a time, starting with the weakest opponent, Denmark, then taking on Sweden and finally Russia. But Parker would not sail for the Baltic until after his blasted ball. Nelson and several similarly frustrated brother officers wrote letters of complaint to friends at the Admiralty. The result was a chilling rebuke for Parker from the First Sea Lord, the Earl of St Vincent. Having believed the delay was caused by lack of a fair wind for the Baltic, St Vincent was naturally livid when he learned the truth. In a private letter, the First Sea Lord advised Admiral Parker that any further loitering at Yarmouth because of some 'trifling circumstance' would result in 'irreparable injury' to the Admiral's career. It had the desired effect, for Admiral Parker called off the ball and sent his wife back to London. By 11 March he was aboard HMS *London* and the Baltic Squadron sailed.

His Indecision was Final

In his cabin, the Admiral brooded over what had happened in Yarmouth and bad weather during the voyage north made his outlook even bleaker. The shallow waters of the Baltic were of very great concern, with the worry of ships running aground. And how would the fleet manage to close with, and destroy, a Danish fleet anchored directly under the gun batteries of Copenhagen? Nelson's fear that Admiral Parker was simply too timid for the mission was confirmed when he went aboard the *London* for a conference as the fleet neared the Baltic. Nelson heard Parker declare that it was his intention to take the Baltic Squadron no further than off the Skaw, which was 175 miles from Copenhagen. There he would await orders to proceed further, something that Nelson knew he did not need to do. But Parker was not to be swayed.

While at the Skaw, severe weather almost ran the *London* aground and the turbulent weather, combined with Nelson's persistence in urging a swift attack on the Danes, finally overcame Parker's fears, temporarily at least. As the fleet neared Copenhagen, the Admiral became more depressed by the hour '...his gloom spreading through the *London*'.[4]

Going aboard *London* for another conference, Nelson was not deterred by the depressed atmosphere. He urged Admiral Parker to discard caution and even proposed ignoring Copenhagen to attack and destroy the Russians at Reval. Parker feared the British would suffer severe casualties and then have to fight their way back out of the Baltic past an undiminished Danish fleet. For once the old man made sense and Nelson agreed that perhaps the Danes had to be dealt with first.

But, as the British drew level with the Koll Peninsula in Sweden, Parker's raging caution was fuelled by the British Government. Its representatives were making renewed efforts to reach a negotiated settlement with the Danes that would see them withdraw from the league and agree to their ships being inspected. These overtures failed to bear any fruit, so it was time for action. Aboard *London*, for yet another command conference with Parker, Nelson was disgusted that the best plan that could be devised was to loiter menacingly just beyond the narrow approaches to Copenhagen. After this meeting, Nelson wrote to a friend: 'I wanted to get at an enemy as soon as possible and strike a home stroke.'

At meetings aboard HMS *London* on 24 and 25 March, Parker is said to have realized that the best way to resolve the need for decisive action with his complete inability to deliver it, was to cede control to Nelson.[5] Here was a dynamic young man who had a superb grasp of both strategy and tactics and saw, with frightening clarity, what was needed. 'It was probably the first time a man as cautious as Sir Hyde had ever seen a brain at work on what could be called precision gambling.'[6]

I really do not see the signal.

The nature of the gamble was to go through shallow, treacherous waters under the big guns of the fortifications and block ships to destroy the enemy fleet and, if need be, occupy Copenhagen. It was agreed that Nelson should lead the direct attack on Copenhagen, taking the lighter draught two deckers in close for a point-blank battle, while the *London* and seven other bigger ships would stand off in support. After the Danish fleet had been destroyed at its moorings, bombardment vessels and troops would deliver the final blow to the city. To lead the attack, Nelson shifted his flag from the *St George* to the *Elephant*, one of the two deckers. The Baltic Squadron had departed from Yarmouth on 12 March, arrived off Cronburg on 24 March and began the move against Copenhagen on 30 March. Parker was sure the Danes would destroy Nelson's ships

The notorious signal No. 39 flown from HMS *London* to instruct Nelson to 'Discontinue the action'. *Illustration by Dennis Andrews.*

with their shore batteries and block ships, but Nelson remained confident. What lay ahead was one of the most remarkable contests in naval history. Wooden walled ships can expect to come off worse when pitched against stone fortifications bristling with guns, but then Nelson was notorious for defying convention.

The British ships sailed down the Sound – the narrow channel between Sweden and Denmark – on 1 April and then, with a favourable wind, swung around to make their attack the following day. Three of Nelson's ships of the line soon ran aground on a shoal, but they continued firing. The bigger, better-protected and better-armed British men-of-war put themselves alongside their more puny Danish counterparts and battered them mercilessly, though not without suffering pain themselves. To Parker, some distance from the mayhem, it looked very much like matters were going against the British. HMS *London* was four miles from Nelson's ship, when she signalled, sometime between 1.15 p.m. and 1.30 p.m., 'discontinue the action'. It might seem clear enough what it meant – the signal was made to the entire fleet and left no discretion. Some British ships obeyed the order, and found out

Nelson coming ashore to negotiate the Danes' capitulation while Admiral Parker remains in HMS London. *AJAX.*

that disengaging in such circumstances was difficult. The rest stuck fast, following the lead of Nelson rather than Admiral Parker. Many of the captains had fought alongside Nelson in earlier battles, so they knew he would be determined to see it through.

Some historians believe that Admiral Parker knew the commander on the spot would be obeyed first and foremost, and he only ordered the signal to disengage as a means of providing Nelson with the correct authorization should he choose to retreat. Others are not so kind, interpreting it as the action of an old man who simply lost his nerve. At the same time as the notorious signal was being run up on HMS *London*, her commanding officer, Captain Robert Otway, was arriving aboard the *Elephant*. He had come across in a rowing boat under orders from Admiral Parker to find out exactly what was happening in the thick of the action. He dodged through a hellish fire to get there, but it was obviously a journey made in vain. With the battle turning decisively in favour of the Royal Navy, Nelson was not minded to call off the action. After seeing the signal he is said to have become very agitated, pacing up and down and stating:

You know...I have only one eye – and I have a right to be blind sometimes.

Nelson raised his spyglass to his right eye and declared:

I really do not see the signal.

He added:

Damn the signal.

The Danish fleet was done for, even though British ships were running aground and the bombardment from the guns in the fortifications was devastating. 'Danish resistance began to falter; the burning flag ship blew up... two more ships cut their cables and foundered... .'[7] A ceasefire was agreed shortly thereafter.

On being rowed across to HMS *London* to give his report on the battle to Admiral Parker, Nelson is said to have remarked: 'I have fought contrary to orders, and I shall perhaps be hanged. Never mind. Let them.'[8]

According to Christopher Hibbert, in *Nelson*, even the anxious old Admiral was delighted by the outcome of the battle – he was therefore not about to recommend his second in command should be court-martialled for disobeying his unwise order.

As Nelson and Parker toasted victory in the Admiral's cabin aboard *London*, it was agreed that Nelson should go ashore to conduct negotiations for an extended truce. On 5 May Nelson visited *London* again, this time to assume command of the Baltic Squadron. Parker handed him a letter transferring command, which said:

In pursuance of directions from the Lords Commsrs of the admiralty; You are hereby requested and directed to take all the ships and vessels of this squadron under your Command and act with them agreeable to the instructions you will receive herewith...[9]

Admiral Parker was being recalled to London, the Admiralty's patience with his incompetence having finally been exhausted by the potentially disastrous signal from HMS *London*. 'The old man was reported to be facing a court martial and certainly he saw no more service and died a few years later in quiet retirement.'[10]

HMS *London* prepared herself for action against the Swedes and Russians but there were no more battles for the British squadron in the Baltic that summer, as Tsar Paul had been murdered. The fight went out of the Russians and the League of Armed Neutrality faded away. The *London* returned to the Channel Fleet and was involved in blockading Ushant until the Treaty of Amiens temporarily ended the war between Britain and France. In late 1805 *London* was set to join the fleet off Cadiz, once more under the command of Nelson. Her departure was delayed and so she missed the titanic Battle of Trafalgar where Britain gained domination of the sea for 114 years, but lost Nelson to a French sharpshooter's musket shot.

Chasing Down The Marengo

In March 1806 came the sixth *London*'s most famous single action; the pursuit of the French ship *Marengo*. The 74-gun *Marengo* was a commerce raider that had been preying on British merchant ships in the Indian Ocean and was on her way home to France in company with the frigate *Belle Poule*, a 40-gunner, and with both under the flag of Admiral Linais. The *Marengo* sighted what her lookouts took to be the sails of a merchant convoy off the Canaries in the early hours of 13 March. On getting nearer, the French realized they had made a dreadful mistake. The sails belonged to HMS *London*, the 80-gun HMS *Foudroyant* and 38-gun frigate HMS *Amazon*. Together with five other warships, which were not far behind, they made up a squadron heading for the West Indies. The *London*'s lookouts spotted the

French ships at around 3 a.m. and a sighting report was immediately passed on to the *Foudroyant*, the squadron flagship, carrying Vice Admiral Sir John Borlase Warren.

HMS *London* closed with the two French ships and engaged them, her log recording that Admiral Linais ordered the *Belle Poule* to flee and, a little while later after daylight, instructed the *Marengo* to disengage, as she was taking too much punishment from the *London*. Breaking free was not too difficult as the *London* had been severely damaged herself. But, by late morning, the *Marengo* had been caught, engaged by the 74-gun HMS *Ramillies*, and surrendered. The *Belle Poule* was chased down by the *Amazon* and also struck her colours. The *London* sustained thirty-three casualties – twenty-three wounded and ten dead. The *Marengo* suffered 146 casualties (sixty-three of them killed), with Admiral Linais among the wounded.

The *London*, foreground, chases down the *Marengo* (middle distance). *Illustration by Dennis Andrews.*

After this exciting encounter, HMS *London* joined the Mediterranean Fleet off the Tagus River. In November 1807, the French invaded Portugal and the country's King and Queen, together with other members of the royal family, took refuge with the British fleet. *London* was one of the ships selected to take them across the Atlantic to exile in Brazil, which was still a Portuguese possession. Three years of duty in the Atlantic followed, before the sixth *London* was sent to the breakers in 1811, just five years short of her fiftieth birthday.

Notes

1. David Davies, *Fighting Ships*.
2. Christopher Hibbert, *Nelson, A Personal History*.
3. Ludovic Kennedy, *Nelson and His Captains*.
4. Dudley Pope, *The Great Gamble*.
5. ibid.
6. ibid.
7. David A. Thomas, *Battles and Honours of the Royal Navy*.
8. Robert Southey, *The Life of Horatio*.
9. Letter to Lord Nelson, Vice Admiral of the Blue, from Sir Hyde Parker, given on board HMS *London*, 5 May 1801. Official papers of Vice Admiral Horatio Nelson, National Maritime Museum.
10. Christopher Hibbert, op.cit.

Chapter Five

BOMBARDING SEVASTOPOL

To Steam or Not to Steam?

Although laid down at Chatham in 1827, the seventh HMS *London* did not go into the water for another thirteen years. Officers and men who had served on the previous HMS *London* in the Napoleonic Wars were at her launch, where they saw that she carried a figurehead depicting Queen Victoria in the full flower of her young womanhood. Costing £90,000 to build, the seventh *London* displaced 2,662 tons, was 205 feet long, with a beam of fifty-four feet and carried ninety-two guns. She had a crew of more than 800 officers and men. Having taken so long to get into the water, *London* was not to commission until 1850, when it was proposed she should be converted to a steamship, using engines from other vessels. This idea was abandoned and she remained a pure sailing man-of-war.

The seventh *London* blooded her guns in the Russian War of 1854-1856, a conflict that saw the Royal Navy not only sending substantial numbers of warships to the Black Sea, but also waging war in the Baltic and directly threatening the imperial capital of St Petersburg.

An Anglo-French fleet sailed from Varna, in what is today Bulgaria, on 7 September 1854, and a week later troops were being put ashore on the Crimean Peninsula. The armies then marched on Sevastopol, the principal naval base in the Crimea, beating the Russians at the Battle of the Alma and taking the port of Balaclava. By mid-October, British, French and Turkish troops were ready for a major assault on Sevastopol.

British warships at anchor in a Crimean harbour during the war with Russia of the 1850s.
US Naval History and Heritage Command/NH 123760.

What had brought them to the walls of Sevastopol was a dispute over holy places in Jerusalem. The French had persuaded the Turks, who held sway over the city as part of the Ottoman Empire, that the holy places should come under the jurisdiction of the Roman Catholic Church. The Russians were outraged at the Orthodox Church being excluded and, in May 1853, demanded Tsar Nicholas I be declared the protector of all Christians within the Ottoman Empire. To underscore their determination, the Russians occupied the provinces of Moldavia and Wallachia. Britain came to Turkey's aid diplomatically because of long-standing fears about further Russian expansion south and eastwards that might eventually threaten India. When Russian warships destroyed a Turkish naval squadron at Sinope on 30 November 1853, Anglo-French concern came to the boil, with the two countries sending warships into the Black Sea in January 1854. A defiant Russian invasion of Bulgaria was swiftly followed by declarations of war from Britain and France. The allies decided that the best way to neutralize the threat from the Russians was to seize their power base in the Black Sea while another fleet worked its way along the Gulf of Finland towards St Petersburg.

On 17 October 1854, the *London*, and four other British warships of the Inshore Squadron of the Black Sea Fleet, was tasked with a bombardment mission that would have been familiar to Nelson at Copenhagen - going up against the stone fortifications with wooden walled men-of-war. While the bulk of the Anglo-French naval forces stayed at a safe distance, the Inshore Squadron came within 1,000 yards of the forts that guarded the entrance to Sevastopol harbour, which was blocked by warships scuttled in its channels.

London undertook the task with gun crews badly weakened, as 200 of her sailors had been sent ashore to serve as part of a naval brigade in the trench network that encircled Sevastopol. The depth of the water close to the fortifications had been assessed by sailors in rowing boats, who used cover of darkness to approach as close as they dared.

The naval bombardment started shortly before 11 a.m., with the *Agamemnon, Sanspareil* and *London* moving in after 2 p.m. followed an hour later by the frigate *Arethusa* and 90-gun *Albion*. The *London* was not a steamship and so had to be towed into position by a steamer lashed to her starboard side. The trio dropped anchor off Fort Constantine – their main target – with only a couple of feet of water under their hulls. Men from the towing ship helped man *London*'s guns, enabling the whole of her port broadside to be brought into play.

The British warships soon reduced Fort Constantine's upper batteries to rubble. But guns in the seemingly impregnable lower levels of the fort began to take their toll on the *London* and her companions. The two frigates in particular were sorely battered and set ablaze, pulling out of line to attend to their wounds. The enemy guns next concentrated their fire on HMS *London*. With casualties mounting, four men killed and eighteen wounded, *London* pulled back but, after rallying her gunners, returned to the fight. All the Royal Navy ships came off badly, even though they had knocked out twenty-two of twenty-seven guns in Fort Constantine. The British had not been keen on the bombardment, which the French had wanted as a distraction for a new ground assault that failed in the event, even before the warships started. A correspondent writing for the *London Times* sent back a dispatch that gave a vivid account of the day's action:

...The firing soon became terrific. At the distance of six miles the sustained sound resembled that of a locomotive at full speed, but, of course, the roar was infinitely grander. The day was

a dead calm, so that the smoke hung heavily about both ships and batteries, and frequently prevented either side from seeing anything. From about two till dark (nearly six) the cannonade raged most furiously. Towards four o'clock, Fort Constantine, as well as some of the smaller batteries, slackened somewhat in their fire; but towards dusk, as some of the ships began to haul out, the Russians returned to their guns, and the fire seemed as fierce as ever. There was one explosion just behind Fort Constantine, which appeared to do much damage. At dark, all the ships returned to their anchorage. The change was magical from a hot sun, mist, smoke, explosions, shot, shell, rockets, and the roar of ten thousand guns, to a still, cool, brilliant, starlit sky, looking down upon a glassy sea, reflecting in long tremulous lines the lights at the mast-heads of the ships returning amid profound silence.[1]

The bombardment had been a futile gesture, with forty British sailors killed and nearly 300 wounded and no sign of Russian resistance crumbling.

Mother Nature was to add insult to injury when HMS *London* and other warships of the British fleet were nearly destroyed by a violent storm that ripped along the coast of the Crimea the following month. The *London* suffered serious damage, while more than thirty transport vessels were lost, with a number of warships.

London's naval brigade sailors would endure the misery of the trenches until Sevastopol fell in September 1855, after which they rejoined their ship. She was returning home to prepare for a new campaign in the Baltic that would aim to destroy the Russian naval fortress of Kronstadt and open the way for an assault on St Petersburg. To prepare her for the job ahead, *London's* guns were removed as her role would be '...to replace mortar batteries and beds, supply shells and provide engineering support.'[2] However, before the massive British assault could be unleashed the war ended. The Russians agreed to humiliating peace terms because they feared their beautiful capital city would be reduced to rubble, and the Austrians were threatening to join the war on the side of the allies. Russia was also on the verge of bankruptcy. To mark the end of the hostilities, the Royal Navy staged a grand Fleet Review for Queen Victoria at Spithead in April 1856, with HMS *London* taking her place as one of two dozen battleships on show.

At War With the Slave Traders

The Russian War had proved that steam propulsion could be a definite advantage and, even though their sails and masts were retained, many British warships, including HMS *London*, were rebuilt to accommodate it.

In January 1857 she arrived at Devonport Dockyard in Plymouth to be converted into a steam powered screw ship.

The London *was lengthened and fitted with new machinery, and her performance was so outstanding that she was used as the model for the subsequent conversion of* Rodney *and the* Caledonia Class *three-deckers.*[3]

London's initial tonnage had been 2,590 tons, but the conversion added a further thirty-six tons. During the rebuild, which took until the summer of 1858 to complete, *London's* guns were reduced to seventy-eight, but they were new, bigger calibre weapons. Sea trials off Plymouth in late 1858 were a success and she proved to be a good sailing ship as well as an excellent gunnery platform. In 1859 the *London* joined the Mediterranean Fleet and in 1863 was paid off into reserve, where she would languish until a crusade against one of mankind's worst evils called her back into service. The British had passed an act abolishing

The *London* shortly after arriving at Zanzibar to act as flagship for the anti-slaving campaign.
Illustration by Dennis Andrews.

the slave trade in 1807 and four years later made it punishable as a felony, but this in itself was not enough.

> *...it was not to be expected that so profitable a trade as the round voyage carrying piece goods to Africa, black slaves to America and the West Indies, and sugar, tobacco, and rum back to Europe, would stop overnight.*
>
> *Before the acts, British ships had carried a large proportion of all the slaves taken across the notorious Middle Passage, and this may have stirred the national conscience of Britain to prosecute the long and thankless war against the slavers, on both the west and east coasts of Africa.*[4]

Thanks to the efforts of the Royal Navy, which was ordered to wage war against the slavers, the trade on the west coast of Africa had been largely stamped out by the 1870s. Unfortunately, it remained active on the east coast where HMS *London* was to play a key part in stamping it out.

Commissioned back into service and sent to Zanzibar as flagship of the Royal Navy's east coast anti-slaving forces in 1874, *London* was soon sending out boats packed with well-armed and vigilant sailors to catch the slave traders.

The journal of an anonymous lieutenant serving aboard HMS *London* noted that in 1875 thirty suspect dhows were intercepted, twenty-six in 1876 and twenty-five in 1877.[5] Stopping and searching the dhows of the slave traders was a treacherous task. During one such incident, the lieutenant in charge of a stop and search party from the *London* would have been murdered but for the intervention of his quick-witted interpreter.

> *The slavers intended to allow him to board, and then to shoot him; but the officer was saved by the interpreter, who, catching sight of a half-hidden Arab, with his gun cocked and levelled, gave warning of the danger.*[6]

The anonymous lieutenant serving aboard *London* noted in his journal on 15 May 1876:

> *This morning the Sultan burned a dhow that his soldiers had captured in the act of carrying slaves after his proclamation.*

The entry for 16 May reported:

> *Painting masts and yards...4-30 steam pinnace returned (with 25 slavers) in charge of Lieut O' Neill, having captured and destroyed two dhows, the remaining slavers and the crews having escaped into the bush.*

The following day the slavers were landed for trial. The cases were not necessarily cut and dried. If it turned out that the owners of the dhows were not guilty of slaving, the vessels were restored to them along with any cargo. In such a case, as the *London*'s lieutenant noted in a list of captured dhows, the verdict could be 'restitution of dhow, liberations of slaves'.

While there were violent incidents in which lives could be lost, life on the Zanzibar station in HMS *London* could be mind-numbingly dull, with nothing much for the lieutenant to write in his journal. The entry for Wednesday, 24 January 1877 reveals 'painting masts and yards'.

This was followed by:

> *Friday. 26. Paint ship's side. Saturday. 27. Painting ship's side. Exercised fire quarters.*

A bit of excitement was provided on 24 May that year by Queen Victoria's birthday, which enabled the *London* and other British ships on station to dress themselves in bunting and fire 21-gun salutes.

HMS *London* contained not only prison cells for holding slavers before their trials but also a small hospital to provide treatment to any sailors that succumbed to the many diseases present in the tropical climate. Several of her officers and a number of her ratings were invalided home, while others ended up dead. The lieutenant noted in his journal on 6 September 1877 that a funeral party was sent to a nearby island 'to inter the remains of the late Samuel Cunningham, Bandsman'. By the summer of 1878, the flagship's commanding officer was Captain Hamilton Earle and he continued to make sure that the *London*'s boats were very active. One patrol commander was Petty Officer Cornelius Duggan who, one night, took a boat out to picket a small island notorious for slave trading.

> *In the small hours, two canoes full of people suddenly quitted the small island... Duggan instantly gave chase. The Arabs opened fire, and several bullets struck his dinghy, while one passed through Duggan's clothes. The pursuit was, however, most pluckily persisted in, until one of the dinghy's oars broke; whereupon Duggan and his companion had to content themselves with emptying their revolvers after the fugitives...*[7]

In the summer of 1880, Captain Charles Brownrigg was appointed as commanding officer of the *London*. On 3 December Captain Brownrigg was killed while leading a boarding party of ten men out on patrol in *London*'s steam pinnace. They came across a dubious looking dhow flying the French flag and it was while Captain Brownrigg was inspecting the papers of its twenty-five Arab occupants that violence erupted. The Arabs fired a volley of musket shots into the British boat and then boarded her. A vicious hand-to-hand struggle developed, in which the British sailors came off worst and were thrown overboard or killed. Captain Brownrigg, who had managed to pick up a rifle, killed one of the suspected slavers with a shot and then, using the rifle as a club, attempted to take some more of them with him.

He was shot a number of times and chopped down but still wouldn't give up until another bullet killed him. Only four British sailors avoided injury or death and, after the dhow had

The *London* at Zanzibar, around 1884. *US Naval History and Heritage Command.*

disappeared, they climbed back aboard the pinnace and returned to HMS *London*. Despite such setbacks, the Royal Navy persevered in its war against the slavers and, although the evil trade didn't disappear completely on the east coast of Africa until the early years of the twentieth century, it was at least sent into a state of terminal decline. The seventh HMS *London* ended her days in Zanzibar, being broken up on station in 1884.

Notes

1. Quoted in *The Royal Navy, a History Volume Six* (published 1901) by William Laird Clowes.
2. Essay by Andrew Lambert in *Seapower Ashore, 200 years of Royal Navy Operations on Land,* edited by Captain Peter Hore RN.
3. Andrew Lambert, *Battleships in Transition.*
4. *The Oxford Illustrated History of the Royal Navy.*
5. Journals kept by an anonymous lieutenant in HMS *Iron Duke, London, Kite* and *Valiant,* and the Chinese gunboat *Eta* on passage to the East, 1874-79. Royal United Services Institution, Naval manuscrips/National Maritime Museum.
6. William Laird Clowes, *The Royal Navy, a History.*
7. ibid.

Chapter Six

AT WAR WITH THE KAISER

Going to War in an Obsolete Fleet

When the half-English Kaiser Wilhelm II came to the German throne in 1888, his admiration and envy of the Royal Navy, and the way in which its warships ensured the security of Queen Victoria's far-flung empire, prompted him to see the building of a German battle fleet as a top priority. France and Russia were also traditional naval rivals, but neither was as muscular or financially robust as Germany, which dominated the Continent via its superb army and its burgeoning industrial might.

But the Royal Navy dominated the seas, governed by the famed Two-Power Standard that required it to be able to match the power of the next two biggest fleets. In 1895 the Royal Navy had thirty-one major warships, which equalled the combined total of those belonging to France, Russia and Germany. In that year iron hulled warships of 15,000 tons displacement with triple expansion steam engines, powering two propellers and capable of more than sixteen knots, were joining the Royal Navy's front line battle squadrons. These vessels were the nine battleships of the Majestic Class. Their armour belt was nine inches thick, they were armed with four 12-inch breech-loading rifled guns in turrets (to batter enemy battleships), twelve 6-inch guns in casemated mountings (to tackle cruisers), six 2-pounder quick firing guns and twelve 3-pounders (to counter the growing menace of small torpedo boats). The Majestic Class was the pattern for the next five classes of British battleship – Canopus, Formidable, London, Queen and Duncan. The Londons (*London*, *Bulwark* and *Venerable*) were the same as the Formidables (*Formidable*, *Irresistible* and *Implacable*) except the former had the advantage of specially toughened Krupp steel armour plate.

The pre-dreadnought battleship HMS *London* is launched at Portsmouth in September 1899.
US Naval History and Heritage Command.

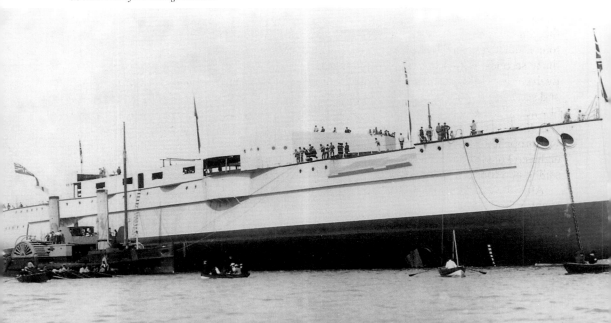

The *London* not long after being commissioned into service. *US Naval History and Heritage Command.*

All three *Londons* were completed in 1902 and the Queen Class (HMS *Queen* and HMS *Prince of Wales*) finished two years later, were almost identical. The Formidables, Londons and Queens, '...constituted a homogenous group of eight first-class battleships, which may be regarded as the best heavy fighting vessels that British naval architects and shipbuilders of the 19th century were capable of producing.'[1]

The eighth HMS *London* cost £1,031,355 to construct and was launched on 21 September 1899 at Portsmouth. Her crew numbered 780, her deep load displacement was 15,000 tons and her armoured belt of Krupp steel was nine inches thick (the equivalent of twenty-seven inches of wrought iron armour). It was originally suggested that the Londons should have an eight-inch armoured belt, as this would save weight and also money (some £47,000) but that idea was abandoned as a false economy.[2] For her main armament, HMS *London* carried four 12-inch guns in two twin mountings, one fore and one aft. They could fire a shell every thirty seconds, with a velocity of 2,562 feet per second. *London*'s secondary armament was twelve 6-inch guns, which could fire a shell every six seconds. She had sixteen 2-pounder and six 3-pounder guns plus four 18-inch torpedo tubes and a pair of Maxim machine guns. She normally carried eighty rounds per 12-inch gun and twenty rounds per 6-inch gun. She was 431 feet in length, with a beam of seventy-five feet and draught of twenty-five feet and was propelled by two sets of triple expansion engines turning two shafts. Her top speed was eighteen knots and her twenty boilers used more than eight tons of coal an hour (if she was sailing at a constant speed of fifteen knots).

After being flagship of the Coronation Review Fleet at Spithead in July 1902, an event held to mark the crowning of King Edward VII, HMS *London* went out to the Mediterranean for five years. She then returned to the Channel Fleet, but was obsolete by the time she was back in British waters. She had been made so, not by a new French,

HMS *London* alongside at Devonport in the early 1900s. *Strathdee Collection.*

Russian or German warship but by the latest addition to the Royal Navy – the revolutionary HMS *Dreadnought*.

Completed in December 1906, *Dreadnought* displaced 17,900 tons and could do nearly twenty-one knots but, most impressive of all, was her ten 12-inch guns in five twin mountings. *Dreadnought* was fast, well protected, heavily armed and, with her main guns all controlled by a single officer in a control top, the full weight of her fire could be concentrated easily. Her striking lines were, literally, the shape of things to come in all of the world's navies. Most important of all, *Dreadnought*'s debut signalled the beginning of a new battleship construction race between Britain and Germany that was to be a major cause of the First World War. While *London* and the other pre-dreadnought battleships found themselves relegated to the second division by 1910, they were still useful for cutting edge aviation development work in 1911-12. A wooden platform for launching aircraft was constructed and was, in turn, fitted to *Africa* and *Hibernia* before going to the *London*. The *Africa*'s flight took place while she was moored to a buoy at Sheerness and the aircraft needed 100 feet of trackway before taking off from the forecastle. When the same aircraft took off from *Hibernia* it needed sixty-five feet less to take off, but the battleship was steaming at around ten knots.

The flight from HMS *London* also took place from the forecastle, but while the battleship was steaming at twelve knots. The wind was from right ahead and Force Three, enabling a run of only twenty-five feet before the aircraft was in the air.[3]

Young Midshipman Drage

On 28 June 1914 Archduke Franz Ferdinand, the heir to the Austrian throne, was assassinated in Sarajevo. The Austrians believed the Serbians were responsible and, already fearful of Belgrade's expansionism, felt compelled to consider declaring war. Germany pushed the Austrians to do so, while the Russians decided they had to assist their Slav cousins in Serbia if conflict came. Britain feared war would spread, regarding the Germans as eager to assert themselves as masters of Europe and reinforce their overseas empire.

Midshipman Charles Drage.
C.H. Drage Collection.

On 30 July seventeen-year-old Cadet Charles Drage woke to find a telegram on his morning tea tray. It was from the Admiralty and said:

Cadet Drage. Join Naval Barracks Devonport immediately.

He leapt out of bed, had a quick wash and pulled his uniform on. '...my kit was hurriedly crammed into my kit-bag. I said good-bye, I am afraid in the highest spirits, and caught the first train to Plymouth.'[4]

The train from London to the south-west was packed with naval officers and ratings and at each station more squeezed aboard. At 6 p.m., Cadet Drage arrived at the barracks, expecting to be sent to the cruiser HMS *Cornwall*, on which he had recently spent time as part of his naval training.

Instead he joined the old pre-dreadnought battleship HMS *Majestic*. He was disappointed to find her boilers were being replaced, so she was not likely to sail in the near future.

The telegram that Midshipman Drage received on 30 July 1914, ordering him to Devonport. *C.H. Drage Collection.*

War between Britain and the Central Powers was declared on 4 August when Germany invaded Belgium. Britain had underwritten Belgian independence since the 1830s, and so was bound by an agreement to send its army to the Continent to repel the invaders. In Devonport, the captain of the *Majestic* gathered his crew together and told them to redouble their efforts to get the ship 'in fighting trim'.[5] However, when it became clear *Majestic* would be in the dockyard for a lot longer, her newly appointed midshipmen were dispersed to other warships, with some going to the new dreadnought HMS *Erin* and three, including Midshipman Drage, being sent to HMS *London*.

London's Coat of Many Colours

The three midshipmen joined HMS *Queen* at Portsmouth on 23 August, then sailed with her to Spithead and, after spending a night in HMS *Irresistible*, finally reached HMS *London* on the morning of 24 August. *Queen, Irresistible* and *London* were part of the Channel Fleet, a collection of obsolete battleships kept on the south coast while the modern dreadnoughts and super dreadnoughts of the Grand Fleet were based at anchorages in Scotland, where they waited for an opportunity to lock horns with Germany's High Sea Fleet. The Channel Fleet's most recent job had been to provide cover for the vessels conveying the British Expeditionary Force (BEF) across the Channel. Crewed mainly by inexperienced sailors from the Royal Naval Reserve, the Channel Fleet warships were of doubtful value and some thought them a liability. To make them less vulnerable, great efforts were put into devising new paint schemes for the fleet's warships, including HMS *London*. Midshipman Drage noted in his diary:

> *At first the hull was painted dark grey and the turrets and upperworks light grey. The idea of this was to increase the difficulty of range-finding by making it impossible to get a 'good cut'. However a moderately accurate range could still be taken by means of the masts or the bows and so the ship was then painted in white stripes and splodges, giving the effect at long-ranges of an unsolved jig-saw puzzle.*

By 28 August, with the realization that florid paint schemes made little difference, the *London*'s startling colour scheme had been covered over by standard warship grey. Despite the vulnerabilities of the Channel Fleet's ships, Royal Navy confidence was high and the captain of HMS *London* gave his sailors a pep talk to bolster them for the struggle ahead. He told them German guns and machinery were inferior to the British fleet's as were the enemy's sailors. While their discipline was good, the *London*'s captain believed the German matelots would 'break down in a prolonged action'. So that his men did not get over confident, the captain warned them the Royal Navy could expect 'some hard knocks'.[6]

Spy Fever

On 29 August leave was granted to *London*'s crew while the battleship was alongside at Portsmouth. Some of the men got extremely drunk and in one instance this led to a shocking allegation.

> *...an accusation of espionage was brought by a Stoker R.N.R. against another of the same rating...*

explained Midshipman Drage in his diary entry for that day.

> *Both men were palpably under the influence of drink but it was apparent that the alleged spy had been acting in a very suspicious manner and he was placed under arrest.*

A few days later the suspect left the ship and no more was heard of him.

By early September the Channel Fleet was at Portland, the broad Dorset harbour dominated by the towering rock of Portland Bill. The fleet's job was to wait for any bid by the High Sea Fleet to break south and it was meant to hold up the Germans long enough for the Grand Fleet to destroy them. There were tactical manoeuvres and small bore gunnery exercises, but a lot of time was spent at anchor, with the crews of the warships being sent ashore frequently for invigorating route marches.

On 22 September disaster struck, with the Sheerness-based cruisers *Cressy, Hogue* and *Aboukir* sunk in the southern North Sea by torpedoes from a single submarine. The Channel Fleet was not to be deterred by this disaster, putting to sea the following day for sub-calibre gunnery exercises. Midshipman Drage noted in his diary:

> *...we cut the* Implacable *target adrift – to their great disgust – and both our turrets missfired, owing to the three quarter charges being rammed too far home.*

Some of the misfires were caused by the inexperienced reservist gunners failing to press the triggers hard enough. The *London*'s main guns also suffered from hang fires, where the trigger was correctly pressed, but there was a time lag before the weapon fired.

Paranoia over German spies was well embedded in the Channel Fleet, with stories about pigeons carrying messages across the Channel every time the warships sailed. There was also concern about secret agents eavesdropping on officers' careless talk in pubs and hotels ashore. The captain of HMS *London* gathered his officers on the quarterdeck and 'warned them about the leakage of information taking place and the great dangers attendant on it'.[7]

On 25 September the fleet was to sail for the Medway to plug the gap left by the loss of the three old cruisers. *London* was to occupy a berth at Sheerness previously taken by HMS *Hogue*, but this move was cancelled, which was perhaps just as well.

> *...we had been ordered to take up the Patrol on which the* Cressy, Hogue *and* Aboukir *had been sunk, regardless of the fact that we were at least three knots slower than those unfortunate ships. Luckily saner counsels prevailed...*[8]

A Sister Ship Explodes

On 26 November, calamity hit the Channel Fleet as Midshipman Drage recorded in his diary:

> *At 8.00 a.m. H.M.S.* Bulwark *blew up, being completely destroyed by the explosion. When the catastrophe occurred I was reading a signal exercise on the port boat deck and had my back turned to the* Bulwark *who was our next astern. I experienced a slight shock, coupled with a blast of hot air and, on turning, saw a vast flame as high as the main truck, around which thick smoke was already beginning to form. Such debris as was in the air consisted of small objects and appeared to be largely composed of wood stored on the booms. There were two distinct explosions and then debris began to fall on our port quarter, a strong wind blowing it away from us. The place where the* Bulwark *had lain was entirely covered with smoke and it was impossible to ascertain the nature, extent or cause of the damage.*

Midshipman Drage was ordered to take one of *London*'s cutters to see if he could pick up survivors and found his boat was first to reach the wreckage:

> *Amongst this debris were about twenty men, some of whom were alive and calling for help. Three of these we picked up. The first man (a leading seaman in the 'gunner's party') was suffering from a scalp wound and broken ribs, the second was unhurt save for burns and the third had a broken leg.*

A column of smoke is all that remains of the pre-dreadnought battleship HMS *Bulwark*.
C.H. Drage Collection.

It took three men to lift each survivor into the cutter and one rescued man was entangled in the wreckage so his sea boots had to be removed to release him. Other boats soon arrived to pick up the wounded and the dying. Only two corpses were spotted, the rest having been carried away by the strong current, but one badly injured man drowned right before Midshipman Drage's eyes. An officer from *Bulwark*, taken aboard the *London*, was given artificial respiration in a desperate last attempt to save him. When they turned him over, *London*'s sailors found a large bolt sticking out of the back of his head. A Royal Marine sergeant brought aboard *London* said he had been eating his breakfast in one of the *Bulwark*'s casemates and the last thing he remembered was seeing flames coming out of the ammunition hoist. Most of the survivors were stokers who had been off watch on the upper deck smoking cigarettes.

A court of inquiry was convened aboard HMS *London* on 27 November and Midshipman Drage was called as a witness. The court sat again the next day and felt able to reach a conclusion. Midshipman Drage told his diary that evening:

> The finding was of course secret but I don't think it was very satisfactory and I imagine inclined towards 'negligence' as opposed to 'enemy agents'.

It was known that ammunition was being shifted around the ship at the time of the explosion and that some of it may have been left lying dangerously exposed in passages:

> The explosion could have been started in various ways, such as a wire fusing, a smouldering match dropped down an ammunition hoist from a casemate, or even a heavy object falling on the nose of a 6" lyddite shell, though the last is extremely unlikely. Once started, it would have spread by the ammunition passages to the cross passages and 12" magazines and touched off every explosive in the ship.

The *Bulwark* had been due to land most of her crew for a route march on the morning of the explosion and, if the steamer had not been late, most of her sailors would have been ashore.

In the late afternoon of 28 November, Midshipman Drage was sent with a party of sailors to search for wreckage and found *Bulwark*'s wardroom dice box and a stoker's ditty box. On 2 December, *London* sent sailors out again to search for more wreckage that might have been washed up. They found a blank German passport – but, rather than being evidence that an enemy saboteur had secreted himself aboard HMS *Bulwark*, it was the usual specimen item carried by British warships so they knew what to look out for. A spy alert was sounded on 3 December when a man was spotted acting in what appeared to be a suspicious manner. HMS *London* sent a boat load of sailors armed to the teeth to investigate, but they found he was a local oyster man. A gruesome task came Midshipman Drage's way that evening, when he was ordered to take some sailors in a cutter to recover a corpse that had been washed up.

> *The body was badly disfigured and in an advanced state of decomposition. Sewing it up in canvas on the mud and by moonlight was a somewhat gruesome business and I was thoroughly glad to leave it on board the* Prince of Wales.

The Sinking of HMS *Formidable*

On 30 December the Channel Fleet put to sea, as it was to head south for gunnery practice off Portland. Three ships – *Exmouth, Duncan* and *Albermarle* – had been detached and were at Portland, and they passed the rest of the fleet coming south as they headed north towards the Medway. By New Year's Eve the weather was worsening as Midshipman Drage recorded in his diary:

> *It was my middle watch and I arrived on deck violently seasick... a wave swept over the foreshelter deck where I was on watch, carried away the canvas screen, nearly dismounted a 12-pounder and injured a man. I found myself under a pile of 12-pounder projectiles...*

The bulk of the Channel Fleet had been kept at Sheerness while plans for a possible bombardment mission against the Belgian and German coasts were discussed at the Admiralty. In late December it was decided that the pre-dreadnoughts needed to get some gunnery practice if they were to stand a chance of being fighting fit for the possible task ahead, hence the decision to send *London* and the other ships to slightly less dangerous waters off Dorset. Hard-charging Acting Vice Admiral Sir Lewis Bayly had been made commanding officer of the Channel Fleet in mid-December 1914 and he was keen as mustard for some energetic gunnery and tactical manoeuvring practice.

The fact that no warnings of submarine activity in the Channel had been posted encouraged him to think he could carry out some leisurely tactical exercises as his fleet sailed south. An escort of six destroyers was provided between the Nore and Folkestone, but thereafter the Channel Fleet battlewagons were on their own. By the morning of 31 December the fleet was only thirteen miles south of Portland, but soon turned back on its course to continue the exercises.

The Channel Fleet exercised throughout the day on New Year's Eve, the seas having calmed down. Midshipman Drage later told his diary:

> *At 7 p.m. we altered course 16 pts. and soon after we exchanged stations with* Formidable... *At 2.20 a.m. a tremendous shock was felt all over the ship... At the same time the lookouts aft saw a flash on the waterline at the* Formidable's *starboard bow and a great column of water was flung up into the air. She had evidently been struck by a torpedo and hauled out of line to port, heeling over slightly, while the* Diamond *and* Topaz *followed to render assistance... At 3.30 a.m. gunfire and Very lights were observed off the port bow, the 'Alarm' was sounded and all the gun's crew closed up. We then altered course for Portland and increased speed to seventeen knots, anchoring at 9 a.m.*

Topaz arrived not long after, carrying forty-three survivors from the *Formidable*, with *Diamond* arriving in the afternoon of New Year's Day with a further thirty-five. Midshipman Drage gave this opinion of the *Formidable's* sinking:

> *There is no doubt that we were the object of a well planned and skillfully executed attack by submarines, which attack was made possible by our steaming repeatedly over the same ground, in single line ahead, at very slow speed (8 knots), and without destroyer escort, on a clear moonlight night with a slight swell. Whether we were also struck or not formed a topic of frequent discussion.*

The shock felt, especially in *London*'s No. 4 coal bunker seemed to support the theory, but the battleship had suffered no perceivable damage. In fact *U-24* had fired a torpedo at HMS *Queen* and missed, then let loose two at *Formidable*, with one of them striking home. As the two cruisers went to *Formidable*'s aid, *U-24* fired another torpedo to finish her off, with the old battleship turning turtle before sliding beneath the waves at around 4.40 a.m. Of her 780 crew only 223 survived. On 2 January, Midshipman Drage noted: 'Our divers went down but were unable to find any damage done to the hull by the torpedo we thought struck us.'

The following day the *London* said goodbye to a particular member of her company without much regret:

> *Our padre, the Rev. Miller, has left the ship... His habit of writing letters in Danish and of sending telegrams whenever the ship went to sea, his remarks on the blowing up of the* Bulwark *and other matters, and finally, his being arrested by the sentry in the vicinity of the magazines have combined to render his absence desirable. My own opinion is that he is merely mad, but the lower deck have no doubts on the matter and stoutly assert that when* Formidable *was torpedoed he stuck his head out of his scuttle and shouted to the submarine 'Not this ship, but the next'.*[9]

On 17 January there came a change of command as a result of the loss of *Formidable*. Admiral Bayly went, to be replaced by Vice Admiral Bethel. The *Formidable*'s sinking had delivered a mortal blow to Bayly's career and ended any notions of sending the elderly battleships up against the shore batteries of Belgium and Germany. The Channel Fleet's warships were expendable but there was no point in making the task of sinking them too easy. It was one thing to lose ships in far away waters, but to lose them at a prodigous rate in home seas would look like sheer stupidity. By early March, the Channel Fleet was much depleted, due to ships being dispatched for duties elsewhere, and the anchorage at Portland was a lonely place. But events far to the south in the Dardanelle Straits were to draw HMS *London* into the maelstrom of one of the biggest disasters of the war.

Notes

1. William Laird Clowes, *The Royal Navy, A History, Volume Seven (1903)*.
2. D.K. Brown, *Warrior to Dreadnought*.
3. *Documents Relating to the Naval Air Service, Volume I 1908-1918*. Edited by Captain S.W. Roskill.
4. Diaries of Commander C.H. Drage RN: Vol: I 1914-1916, Imperial War Museum.
5. ibid.
6. ibid.
7. ibid.
8. ibid.
9. ibid.

Chapter Seven

RUNNING THE DARDANELLES

The Eastern Mediterranean Squadron

The *London* was at sea with *Prince of Wales*, heading for an undisclosed destination, when she received a signal containing dreadful news. Midshipman Drage told his diary on 21 March:

> *...the* Irresistible, Ocean *and* Bouvet *had been sunk by drifting mines and the* Inflexible *and* Gaulois *badly damaged in the operations up the Dardanelles... It is probably to replace the two English ships that we have been sent out.*[1]

During their vain attempts to force the Dardanelles with warships alone, the British were relearning the bitter lesson served up during the bombardment of Sevastopol by the seventh *London* in 1854. Ships against fortifications were not enough on their own.

The Dardanelles were the object of all this attention because they were the best avenue of attack for capturing Constantinople, the Ottoman Empire's capital. The plain truth was that landing troops to take the Gallipoli Peninsula, which dominated the waters of the Dardanelles, was essential if the job was to stand any chance of success.

To the Allies, faced with stalemate on the Western Front, striking up through Turkey seemed to offer a realistic prospect of decisive action, reinforcing the Russians while removing an opponent. Many senior officers in the Royal Navy had not been keen on the venture because of the risk posed by gun batteries and mines in the confined waters of the Dardanelles and their fears had been proved correct. Now, with an assault on the Gallipoli Peninsula proposed, the British Army's top brass were wary of a sideshow draining men and matériel away from the most crucial front. But First Lord of the Admiralty Winston Churchill was a powerful advocate for greater efforts in the Dardanelles and he got his way. In the meantime, as Allied troops were gathered for an amphibious assault, the naval bombardments continued.

On her way to the eastern end of the Mediterranean, *London* stopped at Gibraltar to pick up supplies and ammunition, taking on 900 tons of coal, another thirty-three 12-inch shells

The view from HMS *London* across Mudros harbour in spring 1915. In the background is the old Russian cruiser *Askold* along with some French warships and transports vessels. *C.H. Drage Collection.*

Midshipman Drage's sketch of how HMS *London* demolished the Cape Helles Lighthouse.
C.H. Drage Collection.

and 165 shells for her 6-inch guns. At 3.30 p.m. on 29 March she entered Mudros Harbour in the Greek island of Lemnos, which was the main base of operations. Midshipman Drage noted the scene:

> ...*a somewhat desolate spot though an excellent anchorage. The* Inflexible *was burying her men killed in the action of March 18th and the Funeral March, though it sounded inexpressibly beautiful across the water, was not a very cheering welcome.*

And news from other ships in the Eastern Mediterranean Squadron about the resilience of the Turkish defenders was not encouraging either – the forts were seemingly invulnerable and the gunners good.

The Squadron flagship was the new super dreadnought HMS *Queen Elizabeth*, at that time the fastest and most powerfully armed battleship in commission.

On 1 April, the *London*, in company with the *Prince of Wales*, cruised off the Dardanelles, as a precursor to the same two warships entering on a bombardment mission. Minesweepers preceded them, along with a screen of destroyers.

> *The forts on the European side at once opened a hot fire on the small craft, hitting them several times.*

Midshipman Drage told his diary (3 April):

> *We replied with our port 6" guns, whereupon they ceased fire. It seemed extremely likely that that Cape Helles Lighthouse was being used as an observation station for these forts so we turned our attention to this edifice and opened fire on it with a 6" gun, hitting it repeatedly but failing to bring the tower down. By this time the Turks had plucked up courage and opened fire on us from some concealed batteries on the southern side and from howitzers on the north shore. These latter made remarkably good practice and scored the only hit of the action, a shell which struck the rubbing-strake about ten feet aft of Y.3 casemate, but failed to explode and, owing to the extreme steep angle of descent of howitzer shells, did no damage at all.*

> [Later on the battleship] *went to General Quarters again and prepared to give the coup de grace to the much battered Cape Helles Lighthouse...*

> *Having closed to 2,000 yards from the lighthouse, we opened fire with the fore turret. The first two shots missfired and, the third attempt, when the gun finally went off, hung fire, so*

that the turret swung past and the shell only hit the outhouse...However, the second round hit the tower squarely...

At 6 p.m. *London's* guns fell silent, having fired two 12-inch, eighty-nine 6-inch and twelve 12-pounder shells.

On 4 April, Easter Sunday, HMS *London* carried out a run down the Dardanelles. Midshipman Drage recorded:

Action was sounded at 8.15. The batteries on the South shore opened a brisk fire as we entered the Straits, preceded by the Colne *and* Ribble *[destroyers] and followed by the* Prince of Wales *and six trawlers.*

We replied with our 6" guns, firing our first shot at 8.40. Presently we spotted one particular gun emplacement and, as our 6" did not seem to be very effective, gave it five rounds from the fore turret. The last 12" shell burst right in the embrasure and must, so it seemed to us, have blown the gun and crew to Hell. Anyway the gun stopped firing. The trawlers commenced sweeping and, at 9.00 a.m., we opened fire on a body of men on the Northern side. A shrapnel burst over them beautifully and they scattered. In return, a howitzer hit us on the port ash-shoot, but failed to explode and only bent the plate.

By 11.00 a.m. the Turkish counter fire was really hotting up, with guns in forts on both sides of the Dardanelles aiming at *London*. Shells fell all around her and the after funnel was hit, causing a big dent. Bullets holed a steam pipe and punctured boats on the upper deck. The battleship moved further out, beyond range of the smaller calibre Turkish weapons, but still able to hit enemy positions with her own guns. At 2.30 p.m., the *Prince of Wales* took over *London's* position, with the latter's rate of fire subsiding. Thereafter, the only incident of any remark was a shell exploding in the barrel of one of *London's* 6-inch guns. Midshipman Drage explained:

Self-inflicted damage – the 6-inch gun that exploded during *London's* bombardment run up the Dardanelles. *C.H. Drage Collection.*

> *The gun's crew were unhurt but had the fright of their lives... the Gunnery Lieutenant, who was actually sitting on top of the casemate had a truly miraculous escape. Easily the most terrified person, however, was an unfortunate man in the heads [toilets] who had the scuttle blown in on him... .*

Pulling even further out to make good the damage, *London* returned to the gun line at 4.30 p.m. and spent another hour-and-a-half bombarding the shore before *Triumph* and *Majestic* came out to replace her and *Prince of Wales*. Dropping anchor at Tenedos by 8 p.m., a survey of ammunition expended revealed *London* had fired five 12-inch shells and 248 6-inch. It was believed that she had managed to knock out two Turkish guns. Any significant damage suffered was self-inflicted – in addition to the exploding 6-incher, a 12-inch turret had jammed for ten minutes.

The Turks hadn't managed to harm the *London*, but some of her own sailors came close to blowing her apart, as Midshipman Drage revealed:

> *Some genius threw a smouldering match over his shoulder; there being no cordite in the casemate, it naturally fell down the ammunition hoist and a half-charge came up in flames. By the Grace of God it was put out and did not spread but it was very nearly the end of yet another old battleship 'of no military value'. We have no desire to emulate the* Bulwark.

On 11 April *London* and the other ships of the squadron received the following signal, which showed that spy fever was as prevalent as ever:

> *Ships are warned against the activities of an Italian spy named Caputo, who has been trying to obtain work with the Fleet. Distinguishing mark, a scar on the face.*[2]

Landing the Anzacs

Early on the morning of 13 April the *London* and *Prince of Wales* arrived back at Mudros Harbour where they sailed into the midst of an impressive gathering of vessels that included thirty-five transport ships carrying troops.

> *Ship after ship steamed into Mudros Harbour until there were some two hundred of them anchored there, and they made a city on the water... From shore to ship and from ship to ship swarms of motor-boats and cutters ran about. Every vessel flew its flag, the smoke from hundreds of funnels rose up into the sky, and from one direction or another the sound of bugles and military bands was constantly floating across the water. There was movement everywhere... By night thousands of lights and signal lamps sparkled across the bay.*[3]

Preparations for the landings led to feverish activity aboard *London* with her boats rehearsing the landing of troops from the Australian and New Zealand Army Corps (Anzac). Midshipman Drage noted on 16 April:

> *In the evening we embarked about 600 officers and men of the 11th Battalion Australian Infantry from the transport* Suffolk. *When they came on board at dusk they were welcomed, for some inexplicable reason, by the full assembly of the ship's black cats, some half dozen in number, who assembled on the after turret. This ought to be a good omen.*

The excitement grew aboard *London*, which was honoured to be landing a spearhead formation. With less than a week to go, Midshipman Drage witnessed the spectacle of the troops carrying out practice assaults over the beaches of Lemnos:

> *While the troops were manoeuvring ashore, de Wet and I left our boats and climbed a hill close by, from where we had a good view of the beach. On our right the shore was covered with blue-coated French infantry and great, grinning Senegalese, further away on our left Indians*

The Anzac soldiers wait on *London*'s quarterdeck for the start of their voyage to Gallipoli.
C.H. Drage Collection.

were landing mules, while at our feet were lines of long, lean Australians in their broad
brimmed hats and Britishers in huge solar topees, looking rather puny beside the Colonials but
obviously fit and full of fight.

On 22 April *London*'s Captain John Armstrong gathered his officers together in his cabin to explain the task at hand. What lay ahead was the first major amphibious assault of modern warfare, in which *London* was to land A and C companies of the 11th Battalion of the 3rd Australian Brigade, which was being put ashore between Gaba Tepe and Ari Burnu. Of the brigade's 4,000 soldiers, 1,500 of them would travel in the battleships *London, Queen* and *Majestic* (500 soldiers in each ship) while the remainder were to disembark from destroyers. The battleships *Prince of Wales, Triumph* and *Bacchante* were to provide cover with their guns. Though *London*'s officers were familiar with the mechanics of the landings already, thanks to rehearsals, the captain briefly ran through the procedures again. Half a dozen boats from *London* would be used to land one company of troops, split into two groups called the 1st Tow and 2nd Tow. The former consisted of the Launch (containing ninety-six men), 1st Lifeboat (twenty-one men) and 1st Cutter (thirty-five men) while the latter was composed of the Pinnace (fifty-six men), 2nd Cutter (thirty-five men) and 2nd Lifeboat (twenty-one men). An equivalent number of boats from the *Majestic* would assist *London*'s boats in landing the troops of the other company.

The tows of the steam pinnaces and launches would be slipped when the boats were close enough for the troops to pull themselves to shore using muffled oars. The Anzac commander '...wanted a surprise dawn landing with no preliminary naval bombardment, which required that the warships carrying the troops had to anchor in pitch darkness.'[4]

Originally the landings were to take place on 23 April, when it was forecast the moon would set two hours before dawn, allowing plenty of time for the Anzacs to be put ashore under cover of darkness. But, a storm caused a delay of forty-eight hours, and unfortunately the new date for the landings, 25 April, was scheduled to be a brightly lit night, with moonset little more than an hour before dawn.

The slower transportation vessels left on 23 April, with the remainder of the invasion fleet, including *London*, leaving by the following evening. At midday on 24 April two destroyers came alongside *London* to offload the Australian troops and, at 1.30 p.m., the battleship made her way out of Mudros Harbour.

During chats in the battleship's wardroom Australian officers filled in some of the details the *London*'s captain had missed out – the main landing was to be made by the British 29th Division at Cape Helles while the French were landing at Kum Kali and the Royal Naval Division was possibly going ashore at Bulair, the main aim of the last two landings being diversionary. The Australians did not underestimate the challenges they would face at Gaba Tepe, which, as Midshipman Drage noted '...by all accounts bristles with guns and barbed wire'.

At 4.00 p.m. a Divine Service was held on the *London*'s quarterdeck and, for Midshipman Drage, it brought home the gravity of the situation:

> *The troops fell in on the port side and the ship's company on the starboard. It was an impressive scene. After nearly nine months of active service one can realise that this show is going to be no picnic.*

One of the Australian soldiers aboard *London* later recalled that, at half-past-midnight, he and his comrades were woken up 'from our last peaceful slumber and ordered to fall in on the quarter deck...'[5] The soldiers climbed over the side, carefully descending the scrambling nets and into the boats:

> *This was carried out in wonderful precision and silence, each warship unlading its two companies of men at the same time.*[6]

Another Australian soldier aboard *London*, Walter Stagles, remembered:

> *We moved forward to the side of the battleship, ready to go over the side into the tow boat which was waiting below. As we went over the side each man was given a tot of rum...*
>
> *As the boats became full the young midshipmen in charge of the tow boats gave the order to cast off.*[7]

By 2.30 a.m. all the troops had been transferred to the boats. The moon set half an hour later and finally the battleships moved forward at a speed of five knots trailing behind them the pinnaces, which in turn towed the boats loaded with men. Midshipman Drage thought conditions could not have been better:

> *The night had hitherto been clear and moonlit, but now the moon set and it became fairly dark, the sea was calm and conditions seemed ideal.*

Midshipman Drage's sketches of how the warships stood ready off Anzac Cove to disembark troops (top) and how the boats were arranged around *London* to carry the soldiers ashore (bottom). *C.H. Drage Collection.*

The troops were impatient to be on dry land:

> *... a tedious and silent zig-zag course we took for the shore, the moon had just gone down, the great heights of the peninsula were all covered with the early morning spring mist, and consequently the light was ideal for the adventure...* [8]

The pinnaces separated from the battleships and chugged in towards the beach towing the other boats. Aboard HMS *London* Midshipman Drage watched tensely:

> *...there was still no sign of the enemy, it began to seem as if the landing would be unopposed. When the boats were barely a hundred yards from the beach, there was a single rifle shot and, at the same moment, a body of men were seen running along the shore from the left flank... About fifty yards out, the steam boats slipped their tows and went astern. At this moment the enemy opened fire and the pulling boats grounded on the beach under a terrific fusillade from three machine guns and about a thousand rifles.*

Although the *London*'s boats were now very close to the shore, they were not in the right location. Some did not bother to row in, disgorging their troops as soon as they came under fire. One Australian soldier recalled:

> *...a terrific hail of lead enveloped us and cut the water into foam. I threw my oar overboard and got over the side of the cutter at once, for to remain all massed together...was providing an excellent target for old Jacko the Turk... although he from his trenches could not, in that light, see individuals very clearly, he concentrates this fire on the boatloads of men. So it was imperative to 'scatter!' over the side, and into the sea we went.* [9]

Confused recollections of the landing reflect the murderous chaos of leaving *London*'s boats and struggling ashore:

> *Now the rifle fire was intense, four chaps being killed in my boat. Shot by chance hits through the head, these poor fellows lay at the bottom of our boat, being unavoidably trampled by the clambering men, getting over board... here it must be remembered the unwritten law of each man for himself came into force. Well, over the side I went... holding my rifle not too light up in my right for we did not know the depth of the water we were in, and to flounder out of one's depth, with all this tremendous weight of gear, simply meant you would not rise again... .*

Several others were unable to stand and so drowned. But, somehow, he got to shore '... gasping like a bull out of breath'. [10]

Walter Stagles also survived and, after catching his breath, was ready to get in among the enemy. Away they went into the scrub and took cover. The Australians could see the Turks

HMS *London*, in the foreground, bombards Turkish positions above Anzac Cove. *C.H. Drage Collection.*

fleeing into the distance, but it was difficult to organize an effective pursuit as units were hopelessly mixed up. It was the same story across the Gallipoli landing sites. The Allied troops failed to capitalize on the initial surprise and, with the onset of daylight, the enemy increased the intensity of response. Midshipman Drage told his diary:

> *With the daylight, the Turkish artillery fire, which had hitherto been singularly ineffective, at once increased both in volume and accuracy.*

At 4.50 a.m. the forts at Gaba Tepe opened fire, but HMS *Queen* and HMS *Bacchante* silenced them after an hour's firing. The rest of the squadron, including *London*, also battered concealed Turkish guns on the hillsides. At 9.30 a.m. one of the enemy's heavy ships came down to Chanak – either the German-crewed battlecruiser *Goeben,* or the Turkish pre-dreadnought *Haireddin Barberosse* – and opened fire, but it didn't hit anything.

Ashore the casualties were heavy, with hospital ships leaving for Alexandria by dusk packed to bursting point with wounded men. Midshipman Drage recorded:

> *...the question of accommodating the wounded became more and more pressing... Transports were hurriedly turned into hospital ships and the* London *herself took in 79 serious cases, in addition to our own casualties from the boats.*

A Bitter Lesson in the Futility of War

The Turks had swiftly recovered their nerve and, led by the brilliant Mustapha Kemel pushed the Australians and New Zealanders back into a narrow beachhead in what became known as Anzac Cove. Aboard HMS *London* Charles Drage quickly appraised the Anzac position:

> *Although the Turks had guns on both flanks they couldn't actually see Anzac beach and so see how accurate their fire was. They could have blown us off the beach altogether.*[11]

HMS *London* and the other battleships tried their best to support the troops by bombarding the peninsula. But, the battle became increasingly hopeless. Crisis point was reached on the night of 26 April, with Midshipman Drage one of those sent to try and rescue the Anzacs from what appeared a hopeless situation.

> *At midnight we were suddenly roused and told to man our boats again. The Turks had attacked heavily and driven our men back to the last ridge. The battleships, most of them, got underway and closed the shore.*

Midshipman Drage and other young officers followed behind in their small boats, marvelling at the sheer hell bursting around them. 'The sky over the shore was simply white with shrapnel bursts...'

It appeared a dreadful defeat was about to befall the Anzacs less than twenty-four hours after landing:

> *The commanding officer of the troops on shore, at this moment actually reported that they were unable to hold their ground and must be taken off at once but our Admiral – Thursby – refused to do so and ordered them to hold on at all costs; perhaps one of the momentous decisions in the course of war.*[12]

Certainly an evacuation of the Anzacs would have handed the Turks and their German military advisors a crushing victory that would have been a severe blow to the Allied war effort. In hindsight, it merely prolonged the agony of the Anzacs, leading to the creation of the myth that they suffered more than any other contingent involved in the ill-fated enterprise.

The *Queen Elizabeth* arrived at dawn and used her mighty 15-inch guns to subdue the Turks, enabling the situation to stabilize. So far the *London* had suffered only one casualty – a midshipman shot through the stomach while helping to land troops. Midshipman Drage was frequently charged with taking in boats to bring the wounded out of Anzac Cove:

> *The beach was very crowded and the enemy at once began to search it with shrapnel. The whistle-bang-splash is most disconcerting...*

On 27 April, Turkish battleships again lobbed shells over the intervening land into Anzac Cove, their shells falling in ones or twos quite close to the *London*. The battleships were staying close inshore and suffering mounting casualties caused by bullets and shrapnel from both the Turkish naval barrage and artillery, but Midshipman Drage was most horrified by the lack of proper medical facilities either ashore or afloat. His boatloads of wounded from the beach '...often had to go to one transport already full of casualties... It was one of the big carefully concealed scandals of the Gallipoli campaign'.[13]

Most of the hospital ships had inexplicably failed to return from Alexandria and Midshipman Drage was enraged by the situation it created aboard the transports:

> *The wounded in all these transports must be dying like flies, with few doctors, no nurses and practically no conveniences of any kind. I hope someone hangs for it.*

On the morning of 3 May Midshipman Drage's boat took some casualties to one of the few hospital ships that had returned. He was asked if he would like to have something to eat while the wounded were offloaded. As he tucked in, he witnessed HMS *London* nearly being sunk:

> *A very charming hospital nurse took me into breakfast and placed me where I could see, through the scuttle, the old* London *lying peacefully at anchor about half a mile away. Presently came the familiar 'whirr' and 'crash', and two huge splashes sprang up just under the* London's *stern. Everyone displayed the most praiseworthy activity. I could see the Captain and Navigator nipping up on the bridge and the Commander and 1st Lieutenant charging along the fo'c'sle. At this moment, two more big shells landed, this time one each side of the quarterdeck! Realising that the next salvo would almost certainly hit the ship before the anchor could be hove up, the Captain went ahead and 'snubbed' it out of the ground, having the satisfaction of seeing two more shells drop exactly where the ship had been half a minute ago.*

The guns of the warships continued to pound the Turkish artillery emplacements with little effect and at considerable risk to themselves. Many years later Charles Drage would comment on how ineffective the naval bombardment was:

> *The Turkish guns never stopped firing...and of course a ship's a big target to hit, whereas a single gun in a well-dug in place is a very small target.*[14]

But, luck held for the big ships off the beachheads, with the enemy guns scoring very few hits, but then German submarines arrived and sank HMS *Triumph* close inshore, forcing the battleships to retreat to the protection of nearby islands when not carrying out bombardments.

On 15 May, *London* was informed that she was to be withdrawn to Mudros so her crew could enjoy some rest and recreation. Her farewell gesture to Anzac Cove was an attempt by her 12-pounder guns to silence a particularly annoying Turkish gun. Midshipman Drage reported:

As usual the projectiles appeared to pitch exactly where the flash of the gun had been seen and, as usual, the gun continued firing ten minutes after we had stopped. An ancient Turk, clad in blue trousers and a red shirt, was ploughing a field exactly in the line of fire. He did not deign to notice our little attentions and was still ploughing when we left.

The old farmer's indifference to *London's* efforts symbolized the failure of the campaign to make an impression on the course of the war. It ultimately claimed the lives of 63,216 Allied servicemen – 41,148 of them British[15] – for no perceivable gain and the Allied troops were all evacuated by January 1916. The Turks suffered 200,000 casualties and, on home soil, showed the kind of tenacity so lacking in their efforts elsewhere.

The Tedium of Taranto

The *London* was not destined to stay until the bitter end of the Gallipoli campaign, for her withdrawal from the gun line off Anzac Cove was a permanent exit from the Dardanelles. After coaling for the journey ahead, *London* offloaded all her shrapnel shells for other ships to use and, at 5.30 p.m., on 18 May put to sea. On 19 May around noon she came around Cape Matapan in fine, clear weather, managing a constant fourteen knots in lively seas, despite a temperamental boiler. The following day she reached Malta and went into dry dock for some short term maintenance. Midshipman Drage wrote, 'There are strong rumours that we are joining up with the Italian fleet...'

Just under a month earlier Italy had signed a treaty that promised she would come into the war on the Allied side. A key part of this London Treaty was a British promise to contribute a naval squadron to the Adriatic, where Italy's greatest fear was the Austro-Hungarian fleet. Despite being a member of the Triple Alliance with Germany and Austro-Hungary since 1882, rivalry with the Austrians had kept Italy neutral when war broke out. The Austrians held large swathes of territory the Italians saw as theirs. The Italians also felt threatened by Austro-Hungarian possession of the Dalmatian coastline. The Italian Navy received considerable investment pre-war, to compete with the French in the Mediterranean, but, most of all, to counter the Austrian fleet in the Adriatic. The core of the Italian Navy was six new dreadnoughts, carrying 12-inch guns, completed between January 1913 and March 1916, and eight pre-dreadnoughts. The rest of the fleet was composed of eighteen elderly cruisers and a trio of modern cruisers, along with destroyers and motor torpedo vessels. The Italian Navy's main base was Taranto, in the heel of Italy, well placed to keep a watch on the Mediterranean but also, potentially, to seal off the Adriatic.

The Austro-Hungarian fleet was spearheaded by four Tegetthoff Class battleships, armed with 12-inch guns and completed between December 1912 and December 1915. The Austrians also had three effective pre-dreadnoughts of the Radetsky Class and some other old battleships. Their main base was at Pola in the northern Adriatic.

The modern Italian battleships were fine, powerful vessels but the naval high command was timid, hence the request for not only a British squadron (of super dreadnoughts no less) but also a French supporting force. Unfortunately the inexperienced Italian admirals combined their lack of confidence with stubborn pride. They insisted the Allies should put their vessels under Italian command. The French simply refused, while the British were certainly not going to hand over any of their most modern battleships, especially at a time when all available first rank capital vessels were needed to counter the German threat in the North Sea. Therefore the Italians would get what they deserved – a squadron of pre-dreadnoughts. On 21 May Q*ueen, Prince of Wales* and *Implacable* arrived at Malta and docked so they could be prepared to join *London* as the British Adriatic Squadron. Of the likely Italian reaction, Midshipman Drage drily observed: 'What their feelings must have been when this antedeluvian quartet was fobbed off on them can well be imagined.'

On 25 May the *London* received a general signal: 'State of war exists from to-day between Italy and Austria and Italy and Germany. Italy is consequently an Allied power.'[16]

However, Italy was not yet at war with Germany, and would not be so until the end of August 1916. With Italy's hand now revealed, at least as far as the Austrians were concerned, the British Adriatic Squadron set sail for its new operating base, arriving at Taranto on 27 May. Midshipman Drage was impressed with the welcome:

> *The walls were lined by cheering crowds and our reception left nothing to be desired on the score of warmth. It remains to be seen if this enthusiasm will survive war-taxation, reverses and the daily casualty list.*

The joy soon gave way to the jitters, for, in the early hours of the next morning, sentries started firing their weapons wildly in the air and Italian battleships blew their sirens to indicate air attack. It was a false alarm.

A pattern of gunnery shoots in the Gulf of Taranto and route marches ashore was soon established for the British squadron, interspersed with the occasional nerve jangling false alarm. The tedium of Taranto rivalled that of Portland, but at least the weather was more pleasant and the British sailors enjoyed some relaxing picnics. It was so hot that *London*'s sailors often slept on the upper deck.

In early June the British and Italian battleships exercised at sea and Midshipman Drage was not impressed with the seamanship of the new allies:

> *The manoeuvring of the Italians was abominable, and the amount of smoke they make is appalling. The Heavens are literally darkened and it is difficult enough to keep station astern of them, apart from fighting our guns, in the kind of black fog they create. God help us when we have to tackle the Austrian battle fleet.*

London's gunnery exercise results proved variable and her clapped out boilers could never manage more than sixteen knots. The performance of the Italian ships was similarly lacklustre. Midshipman Drage later pondered the likely result of a clash between the motley Anglo-Italian fleet and the Austrians, who were said to be as efficient as the Germans. He regarded the Italians as 'utterly inefficient'.

> *There is little doubt that in an action our Fleet would be defeated.*

In early August *London*'s crew were pleased to hear that the old Turkish battleship *Haireddin Barberosse*, one of the warships that had tried to hit their vessel in the Dardanelles, had been sunk by the British submarine *E11*.

The pre-dreadnought HMS *Duncan* sailed into Taranto on 23 August, to relieve the *London* so she could go for a refit at Gibraltar. On 5 October HMS *London* left Gibralter and nineteen days later arrived back at Taranto. By now young Drage was an Acting Sub Lieutenant but the thrill of promotion was soon dissipated by the usual Taranto tedium. Christmas and New Year passed equally uneventfully but, on 25 February, the Anglo-Italian Fleet was advised to raise steam for possible action. Austrian warships came down as far as Corfu but slipped away before they could be intercepted. The Allied ships didn't even bother to leave harbour.

HMS *London* arrives at Taranto. *C.H. Drage Collection.*

The young officers who lived in HMS *London*'s gunroom between 1914 and 1916. Midshipman Drage is pictured sitting third from the left. Of the others, one young officer went down with *Invincible* at Jutland, two others were killed in action before the end of the war and one drowned when his submarine sank in the early 1920s. *C.H. Drage Collection.*

Austrian Saboteurs Destroy a Battleship

Months passed with nothing significant happening in Taranto. Then, as August 1916 opened, calamity struck in an event recorded vividly by Sub Lieutenant Drage in his diary entry for 2 August:

> *I was sleeping peacefully when I was suddenly awakened by a terrific explosion, followed almost immediately by the noise of alarm rattlers. I rushed up on deck and was confronted by a scene more like those depicted in the illustrated papers than anything I had hitherto seen in the war. The* Leonardo da Vinci *was down by the stern with her quarter-deck practically under water. A terrific fire was blazing aft and there were constant explosions. 'Away all boats!' was piped and Bradshaw (the Navigator) and I scrambled down into the launch, which was being manned with her 'abandon ship' crew of stokers, all full of fight but with the most elementary ideas of rowing. As we got away from the boom, the ships, one after another, switched their searchlights on and the harbour became as light as day. The water was dotted about with swimming Italians who had jumped overboard from the* Leonardo, *where no kind of order seemed to be kept. Crowds of Italian boats were rushing about doing very little good and many more seemed reluctant to close the burning ship. We kept steadily on, picking up survivors as we went. Some were panic-stricken and some were most certainly not so, but one and all joined in making a most infernal shindig, so that, apart from the noise of the explosions, Bradshaw and I and the Cox'n could scarcely make ourselves heard. We steered in for the fo'c'sle where crowds of Italians were still jumping into the ditch. This took us unpleasantly close to the fire itself, where the explosions of the 'ready-use' ammunition were flinging chunks of metal into the water all round. However the end was at hand. Suddenly she heeled and capsized towards us, her foremast narrowly missing the launch while her main mast was equally close to our 1st Cutter with the Captain, Commander and Binney all in her! For one second we looked straight down her fore funnel while the heavy sea slid across her vertical deck. Then she was gone. One solitary sailor ran up her side as she turned, scrambled over the bilge keel and seated himself in the glare of the searchlights on the keel. There was nothing more to be done but to return to our ship... The ship was crowded with badly burnt survivors. I helped the officer of the watch to hoist the seaboats, then made some cocoa and went back to bed.*

The following morning Sub Lieutenant Drage found himself gripped with a dreadful fascination:

> *First thing I saw on looking out of my scuttle was the bottom of the* Leonardo *looking like a stranded whale.*

At breakfast the officers swapped tales of the night before:

> *Everyone full of their experiences last night. People who took survivors straight to Italian ships say that the last vestiges of discipline had completely vanished.*

Whereas *Bulwark*'s destruction had initially been wrongly blamed on sabotage, there was apparently no doubt about what had destroyed the Italian dreadnought. The *Leonardo da Vinci,* only completed in May 1914, was sunk '...by an Austrian sabotage explosion.'[17]

London's crew were extremely happy to hear that she was finally going to sail for home where she was to receive a full refit lasting many months. In preparation for leaving, the ship offloaded fifty 12-inch and 300 6-inch shells which represented her current excess carriage of ammunition. On 26 September the pre-dreadnought *Africa* turned up to relieve *London*.

Sub Lieutenant Drage described his ship's farewell at sea:

Fire in Maintop

Very fierce fire & constant explosions

Crew jumping overboard.

Quarterdeck on fire

Leonardo da Vinci after the explosion.

Sub Lieutenant Drage's sketch of the destruction of the Italian battleship *Leonardo da Vinci.*
C.H. Drage Collection.

We flew our paying-off pendant, a streamer one and a quarter times the length of the ship with a large gilt bladder on the end... At 2.00 p.m. the squadron weighed and formed divisions in line abreast. We steamed down the middle, manning ship and with our band playing 'El Abanico', which is considered 'the London's anthem'. Each ship as we passed cheered us and then, as we got clear of the squadron and passed the booms, all the bands played 'Auld Lang Syne'. One forgot all the heat and boredom and beastliness of the last eighteen months. One only remembered that we had been together for two years of active service and one was damned sorry to be leaving them.

Zig-zagging at fifteen knots, and with an Italian destroyer escort, the London was heading for Malta where she stayed for three weeks. Just before she left a special passenger was delivered aboard. Sub Lieutenant Drage recorded the arrival of the battleship's mystery guest:

At 9.00 a.m., just before we got under way, a harbour launch came alongside and a tall, thin hawk-faced worried-looking man was brought over the side and was

Captain Karl von Müller, who called his time HMS *London* **on the way to a prison camp in England 'unspeakable'.**
US Naval History and Heritage Command/ NH 50785.

received by an escort of marines with fixed bayonets. He was lodged in the Captain's spare cabin... Rumour as to his identity was rife, the lower deck having some truly ingenious theories. However, he turned out to be none other than Captain von Müller of the Emden, *who was being sent back to England in (God knows why) extreme secrecy. He was treated with the greatest consideration. The Captain visited him in person every day to hear any complaints and all his wants were instantly supplied. In fact he borrowed, and never returned, a pair of my trousers and my safety-razor...*

The German light cruiser *Emden* had prowled the Indian Ocean between August and November 1914, sinking nearly twenty merchant vessels and taking the surrender of one Russian warship and sinking another. Her reign of terror was brought to an end by the Australian cruiser HMAS *Sydney*, which trapped the *Emden* off the Cocos Islands, turning the German warship into a burning wreck and forcing her to run aground on a reef. The *Emden's* skipper was being brought back to Britain to be handed over during a prisoner exchange in neutral Holland.

On 11 October HMS *London* reached Gibraltar and left two days later. Bad weather set in as she entered the Bay of Biscay and many of the crew were very seasick, including Sub Lieutenant Drage. But the welcome sight of the Devon coast on 16 October made everyone happier. Sub Lieutenant Drage noted gratefully:

Reached Plymouth at 7 a.m. The whole wardroom came on deck to have a look at England. Captain von Müller was turned over to an officer from the Devons.

As soon as von Müller was safely back in German hands he complained, via the newspapers,

The *London*, dazzle-painted after the refit that converted her into a minelayer. *Topham Picturepoint.*

about dreadful treatment during his time as a prisoner of war, particularly aboard HMS *London*, which he described as 'unspeakable'. It turned out that Prussian pride had been sorely dented by being put aboard the battleship at short notice without baggage, which had to follow in another ship. Borrowing clothing (including Sub Lieutenant Drage's trousers) and toiletries (including Sub Lieutenant Drage's razor) was considered a dreadful indignity by von Müller. But, as the astonished officers of *London* later remarked, he was very well looked after on their warship.

After passing through Plymouth Sound and into the Hamoaze, the *London* entered Devonport dockyard. Sub Lieutenant Drage noted:

> We went into the large basin and secured. Only ships in harbour were the Devonshire *and* Cumberland. *Began paying off the ship for a long refit.*

Between 17 and 19 October the battleship was de-ammunitioned, a 12-inch shell being dropped into the water during this process. Within a week the crew started to disperse, with drafts of men going to various naval barracks.

Sub Lieutenant Drage chronicled *London*'s decommissioning in his entry of 27 October:

> The ship paid off. We said 'good-bye' to the Captain. The hands fell in and marched up to the barracks and the 1st Lieutenant gave the order – 'SHIP'S COMPANY – DISMISS!'

During the refit HMS *London*'s guns were removed as she was being converted to a minelayer but by 1919, the eighth *London* was a depot ship. She was placed on the disposal list in 1920 and sent for breaking up.

Notes

1. Diaries of Commander C.H. Drage RN Vol. I 1914-1916. Imperial War Museum.
2. ibid.
3. Alan Moorhead, *Gallipoli*.
4. *The Oxford Companion to Military History*.
5. Anonymous account of the Anzac Landing, Imperial War Museum.
6. ibid.
7. Walter Stagles, Sound Archives, Imperial War Museum.
8. Anonymous account of the Anzac Landing, Imperial War Museum.
9. ibid.
10. ibid.
11. C.H. Drage, Sound Archives, Imperial War Museum.
12. Diaries of Commander C.H. Drage RN Vol. I 1914-1916. Imperial War Museum.
13. C.H. Drage, Sound Archives, Imperial War Museum.
14. ibid.
15. *Oxford Companion to Military History*.
16. Diaries of Commander C.H. Drage RN Vol. I 1914-1916, Imperial War Museum.
17. *All the World's Battleships*, edited by Ian Sutton.

Chapter Eight

POLICEMEN OF THE EMPIRE

The Treaty Cruiser

The 1921 Washington Treaty conference aimed to prevent war between the major naval powers by restricting warship construction. It decided that cruiser displacement should be limited to a maximum of 10,000 tons per ship. Britain had set the pace by constructing the Hawkins Class – the most modern and capable cruisers of their time – which displaced close to 10,000 tons. It was known that the USA and Japan would soon match, and probably exceed, them, building 10,000 ton cruisers with 8-inch guns (the Hawkins Class ships carried 7.5-inch main guns). The British answer was heavy cruisers of the County Class, constructed in three batches, that were to be armed with 8-inch guns and would be fast with long range. The Washington Treaty left the Royal Navy equal to the US Navy in terms of battleships, but the British fleet's warships were on the whole much older than the US Navy's and they often had smaller calibre guns. The hard running they had endured during the First World War had also left many British ships in poor shape. At the end of 1923 approval was given by the British administration for the building of eight County Class cruisers. But there was a change of government in early 1924, with Labour coming to power and the new administration was inclined to disarm, rather than build up, the fleet. The construction programme was cut, with the number of County Class cruisers reduced to

Cruiser *London* being launched at Portsmouth in September 1927. *US Naval History and Heritage Command.*

five, but, in late 1924, the Conservatives were returned to power and the Royal Navy had its original order restored.

The 8-inch gun was universal for all of the so-called 'treaty cruisers' of the Washington Conference navies. All had 21-inch torpedoes, except for Japan's Myoko Class cruisers, which had 24-inch. The County Class cruisers had armour plate over vital areas, but did not have an armoured belt. The French fleet's Duquesne Class cruisers had no armour at all, but the USA's Pensacolas, Italy's Trentos and Japan's Myokos had armoured belts of between three and four inches. The County Class cruisers were the slowest of the treaty cruisers, with a top speed of about thirty-two knots while the Myoko Class were the fastest, able to do more than thirty-three knots. Both the Myoko and the Trento classes secretly broke the Washington Treaty limit of 10,000 tons – the former, blatantly, at 11,150 tons and the latter having a displacement of 10,510 tons. The Japanese also built Nachi Class cruisers that displaced 10,940 tons. The four Nachis, built between 1924 and 1929, were the most powerful cruisers afloat, being armed with ten 9-inch guns and six 4.7-inch secondary. Japanese and American heavy cruisers had 5-inch secondary armament while the French fitted theirs with 3-inch guns. The Italians matched the 4-inch secondary armament of the British.

First of the counties to be built was the Kent Class (*Kent, Suffolk, Cumberland, Berwick* and *Cornwall*) with two extras (*Australia* and *Canberra*) for the Royal Australian Navy. They were laid down between July 1924 and September 1925 and completed between July 1927 and July 1928. Next came the London Class (*London, Devonshire, Shropshire* and *Sussex*), laid down between March 1926 and February 1927 and completed between January and September 1929.

Like her immediate predecessor, the ninth *London* was built by Portsmouth Royal Dockyard. She was laid down on 23 February 1926, launched on 14 September 1927 and completed on 31 January 1929. The final batch of counties consisted of *Norfolk* (laid down in July 1927 and completed in April 1930) and *Dorsetshire* (laid down in September 1927 and completed in September 1930).

HMS *London* at Gravesend in the late 1920s. *Strathdee Collection.*

The most obvious difference between the London Class and the Kents was the removal of a protective bulge on the waterline, which forced a modified hull design that increased length by two feet eight inches and incorporated internal bulges. Their high freeboard meant they were, in most weather, dry ships and could keep up a good rate of knots even in heavy seas. The beam of the Kents was two feet five inches wider than the Londons. HMS *London's* displacement was 9,850 tons (standard) and 13,315 tons (full load). She was 632 feet 8 inches in length, with a beam of sixty-six feet and a draught of twenty feet nine inches. Propulsion was provided by Parsons geared turbines and eight Admiralty three drum boilers. HMS *London* could carry 3,190 tons of oil and achieve more than thirty-two knots. In addition to eight 8-inch main guns and four 4-inch high angle secondary *London* had four 2-pounder guns and eight 21-inch torpedo tubes. The bridge and the foremast of the London Class cruisers were set further aft, to allow B turret to fire 'abaft the beam without blasting the bridge.'[2] The London Class was fitted with single 4-inch guns because the multi-barrelled 'pom pom' guns were not yet entering service.[1] She carried one catapult launched aircraft and her crew numbered 784.

In June 1927, the League of Nations held a conference in Geneva that aimed to create the mood for more disarmament, primarily through naval construction restrictions. A prime concern was the growth in numbers of cruisers. The British came to the conference with a request for even more, but of a smaller displacement (around 7,500 tons). The Americans wanted to continue building cruisers of 10,000 tons, which is what they needed for patrolling the vast expanses of the Pacific. The cash-strapped British needed cruisers that were smaller and could be used on long-term deployments overseas and so did not necessarily need the same habitability or endurance. The Japanese and Italians had broken all the rules anyway so they pressed for more ships of the maximum displacement. In the end, the 1927 conference collapsed because the British and Americans could not agree tonnage totals for cruisers. However, the London Naval Conference of 1930 resulted in the USA being allowed 180,000 tons of heavy cruisers, the UK 146,000 tons and the Japanese 108,400 tons. When it came to light cruisers the UK was allowed 192,200 tons, the USA 143,500 tons and Japan 100,450 tons.

Trouble in Cyprus

For her first overseas deployment, HMS *London* was sent to the Mediterranean, where she became the flagship of the First Cruiser Squadron.

Trouble was brewing on the island of Cyprus in the early 1930s and soon boiled over into insurrection. It had been granted Crown Colony status in 1925, after being taken from the Turks and occupied by British forces in the First World War. Its position in the eastern Mediterranean made it an ideal base from which to provide further security for British interests in the Middle East, such as the Suez Canal and Palestine. A strategically vital oil pipeline terminated at Haifa and the island was also an important staging post for travellers heading to Britain's empire in Asia. Having cast off the hated Turks, Greek Cypriots were angry and frustrated at being prevented by British rule from uniting with their ethnic homeland of Greece. A decision by the Crown Colony administrators to raise taxes stoked the fires of resentment. On 21 October 1931, a mob gathered and marched on the Governor's residence. They burned it to the ground and then, when the police belatedly intervened, went on the rampage. The Royal Navy was asked to aid the civil powers, with HMS *London* and her sister ship HMS *Shropshire* sent to suppress the uprising in company with the destroyers HMS *Acasta* and HMS *Achates*. Setting sail from Suda Bay, Crete, 500 miles to the west, the British warships steamed at high speed for Cyprus. As flagship, HMS *London* was carrying Rear Admiral J.W.C. Henley and he decided the Royal Navy warships should intimidate the rioters by showing the White Ensign in the island's main harbours. *London* went to Larnaca, *Shropshire* to Limassol, *Acasta* to Paphos and *Achates* to Famagusta.

At Larnaca, HMS *London* landed armed shore parties of Royal Marines and sailors to protect government offices. They found the local population still defiant and capable of violence. The protestors hurled verbal abuse at *London*'s marines and matelots, waving the Greek flag in their faces. Stones rained down on the helmets of the British, but they held firm. The rioters were trying to make the marines and sailors lose their cool and open fire, in the hope that Greece might protest and even intervene militarily. In the early hours of 25 October, the leaders of the revolt were arrested by troops and police. They were taken by boat to *London* and *Shropshire*, where they were held in the convivial surroundings of the cruisers' wardrooms. The detainees were given every hospitality, including good food and drink. A few stiff whiskies helped steady the nerves of those still a little shocked at being hauled from their beds in the wee small hours. The seizing of the leaders took the steam out of the revolt and the British warships were soon able to quit the island's harbours, leaving a job well done behind them.

This widely reported intervention came at an opportune moment for the Royal Navy, which was weathering turbulent times. Defence cutbacks meant it was operating on a shoestring and barely able to keep up its operational commitments. The previous month there had been a mutiny over pay in the Atlantic Fleet and there were widespread, and unjustified, claims that the Royal Navy was turning Bolshevik. The successful intervention by *London* and the other warships in Cyprus proved this was not true.

A Boy's Life
Signalboy Edward Reynolds joined HMS *London* in May 1934 and found the cruiser to be 'one of the best ships in the Navy'. Some years later he recalled that the cruiser was '...a very spacious ship by naval standards of the day, and possibly the most modern ship in the Navy at that time. The County Class were really lovely ships.' Boy sailors aboard HMS *London* were worked very hard:

> We got up on the bridge at six in the morning and scrubbed it down. The people who were supposed to do the fore noon watch finished their scrubbing at seven and went down below, had breakfast and were back on the bridge at eight o'clock to stay there 'till half-past twelve. Then they went down below until half-past one, then came back up and remained on the bridge 'till half-past three. From half-past three 'till four they had their tea and did the first dog watch which was from four 'till six. They'd get two hours off from six to eight and come on the bridge from eight 'till ten. The other watch would do the morning watch from six 'till eight, helping with the scrubbing until seven, go down at eight, back on the bridge again at nine 'till half-past eleven. Half-past eleven 'till twelve, lunch. On the bridge from twelve 'till four, have the first dog off. They'd do the last dog watch and then from eight o'clock 'till the time they'd turned in at nine, they'd be free. Strangely enough, boys who were not on watch had to be in bed by nine on a ship, yet they were allowed to stay up until ten when they were on duty. And one boy slept on the bridge all night in case he was required. This went on until you became a man, so you

had quite a tough day, really. On London *during my time we had one person there, a yeoman, who was quite nasty with the boys.*[3]

The fact that Signalboy Reynolds had a brother on another ship in the fleet, who had made Signal Bosun at a young age, may have made him a special target of jealous spite from the *London's* yeoman.

Turkish Stand-off

When a small boat from the cruiser HMS *Devonshire* was fired upon in the summer of 1934, *London* was among the warships sent to hover with menacing intent in the Aegean. A surgeon lieutenant and a lieutenant had taken the small boat out for a pleasure cruise and strayed too close to a Turkish island. Sentries on the island gestured at the boat, and its occupants thought they were being told to come closer. When they rowed in the Turkish soldiers opened fire, wounding both British officers. Strong diplomatic representations were made in Istanbul and a Royal Navy heavy squadron was sent to the area. Led by the Mediterranean Fleet flagship HMS *Queen Elizabeth*, it included the battleship *Royal Sovereign*, cruisers *London, Shropshire* and *Devonshire* together with destroyers. It emerged that there had been a terrible misunderstanding. To Westerners the sentries appeared to be waving the boat in, but they were actually using a well known Turkish hand gesture to indicate the vessel should go away. In addition, the *Devonshire's* boat had been flying no flags of identification and the two officers were in civilian clothing. Turkey still apologized and paid £2,000 compensation to the wounded men.

Playing Host to Il Duce

With potential enemies such as Germany under Hitler, and Italy under Mussolini, blatantly ignoring the restrictions of treaties to build powerful new fleets, the pressing need for a stronger Royal Navy was obvious. In the short term the Royal Navy's position was not threatened, but as the end of the decade neared it would face serious challenges not only in Europe but also from the Japanese.

But, in July 1935, the Royal Navy felt confident enough to send significant units of the Mediterranean Fleet, including HMS *London*, back to Britain for a Jubilee Review at Spithead in honour of King George V. After the Review, *London* returned to the Mediterranean to resume her role as flagship of the First Cruiser Squadron. The other three cruisers in the squadron were HMS *Shropshire*, HMS *Sussex* and HMS *Devonshire*, but they rarely found themselves working as a unit, for British warships in the Mediterranean were few and the tasks many.

During a visit to Venice in 1935 HMS *London* was honoured with the presence of Mussolini. Signalboy Reynolds watched in amazement as the Italian dictator strutted by, trailing a posse of British naval officers in his wake. The young sailor's lasting impression was that Mussolini was probably,

...ill mannered, but he had terrific amount of drive... and he actually showed our Admiral around the ship rather than the Admiral showing him around. He went where he wanted to go, leading the way, a couple of yards in front of the Admiral, the Flag Lieutenant, the Captain of the ship and the First Lieutenant... Mussolini looked quite purposeful and stern, no smile. I didn't see him speak to anyone. He was very interested in the catapult. We had an Osprey aircraft on board at that time and a steam catapult, which was quite new. The 8-inch guns were also quite modern.

There weren't many other ships in the world with 8-inch guns...6-inch, yes, but not 8-inch. They were good guns. I think they were quite long-range compared with others. Mussolini was very interested in the turret and in the director that was right at the top of the ship.

Corporal Edmund Priestley was a newly married RAF air maintainer posted to 771 Catapult Squadron and sent to HMS *London*. He was sent to sea because shortly after the First World War the RAF had absorbed the Royal Naval Air Service and was therefore responsible for providing warships with their embarked flights. He joined the cruiser at Malta full of apprehension about life on the ocean wave:

I didn't want to go to the Fleet Air Arm at the time... I didn't relish the idea. But, I didn't really get pushed around any more than I would have done elsewhere.[4]

Corporal Priestley mucked in with the sailors and soon got used to the rather basic living conditions: 'Our accommodation was the same as for the sailors. It was a bit crude.'

The five RAF men in the *London* had their own lockers and a big table to sit around for meals. The chairs and tables were folded away at night so they could sling their hammocks.

We had one spot allocated for the RAF personnel and two naval ratings who were attached to us and that was our mess...we used to eat and sleep in the same spot, except in hot weather when we used to sling our hammocks on the upper deck...

Aside from Corporal Priestley, the RAF team consisted of a sergeant, a leading aircraftman metal rigger, a wireless operator mechanic and a fitter/armourer.

And we had a Leading Seaman naval telegraphist who always flew with the aircraft. The pilot was a naval officer. We also had an Able Seaman who was a qualified naval parachute packer. He used to generally help out with anything that was going on to do with the aircraft.

Rescuing the Nuns of Barcelona

The second London Conference of March 1936 was meant to maintain limitations on naval construction, but both Italy and Japan declined to attend.

Japan was proposing that it should be allowed to build a fleet equal in size to those of America and Britain. It saw the conference as a ploy to force it to abandon its expansion plans. The conference achieved nothing and from that point on all naval construction restrictions were lifted. Meanwhile, that April, in the Mediterranean aboard HMS *London*, Signalboy Reynolds was making good his escape from the cruiser's vindictive yeoman: 'I volunteered to go to a destroyer for experience and to get out of his way.'

On 18 July 1936 the Fascist General Franco led an uprising by twelve military garrisons on mainland Spain and five in Spanish Morocco. Most of southern Spain fell to the Fascists and outside forces intervened to stoke the fires of conflict – the Germans and Italians on the Fascist side and Russia on the Republican government side.

Thousands of British passport holders were at the mercy of various warring factions and the UK government responded by sending warships, including HMS *London*, to evacuate them from danger, using Barcelona and Valencia as points of embarkation. HMS *London*'s Paymaster, Commander N. Wright recorded in his diary:

21 July. Discovered at breakfast time that we are being diverted to Gibraltar or vicinity, Spain being in a state of internal revolution. Damnation!!!

The ship stopped at Malta for supplies and then went on to Barcelona in company with the *Devonshire*.

Bombs fall on a Spanish port not far from British warships. *Ballantyne Collection.*

Heading for Barcelona at 29 knots, with instructions to be ready to embark 1,000 refugees tomorrow...

However, the *London*'s crew believed it would all turn out to be 'a damp squib'.[5]

On 22 July the cruiser was secured by her stern to the Mole in Barcelona harbour and the ship received reports from ashore saying the city was now quiet after savage fighting. '...one regiment of Fascists was annihilated in a square, being mown down by concealed rifle fire', Commander Wright told his diary.

...bodies are being burnt today. All churches have been looted and gutted and some shops looted. General shortage of food. Cars driving around laden with folk bristling with rifles.

As *London* waited for the evacuees to start arriving, the situation grew more volatile. On 23 July a bomb from an aircraft landed near a British warship patrolling the Straits of Gibraltar and HMS *London* was informed that Barcelona's British Colony of 1,000 people, including a large number of nuns, was to be evacuated. By midnight on 23 July, only 200 people had turned up, with forty of them being immediately sent to Marseilles aboard the destroyer HMS *Douglas*, which had come alongside *London* to take them off.

One of those who made her escape aboard HMS *Douglas* was the Honourable Mrs J. Larios, the daughter of Lord Ruthven. She kept a diary of the events that led to her escaping via HMS *London*, writing that, on 22 July, someone called 'C' (possibly her husband) managed to get down to the port: '...oh! joy of joys, there's an English cruiser in there tied up'.[6]

The following day she went with her family to seek advice from the British Consul about the best method of approaching the warship, but he forbade them from going to the cruiser.

So we left gloomily and I then said 'to hell with them all, come down to the port and see if we knew anyone on board' ...so we traipsed on another 45 minutes hard walking, boiling sun, down to that lovely sight in the harbour, the 'London' tied up alongside. But oh! Bitter moment... a huge mass of Communists and guns between us and her, and a guard of the Carabineros, who absolutely refused to allow us on board. I could have wept. However, while we were sitting disconsolately on a bollard discussing what to do, an officer came along the pontoon from the ship, and I shouted to him, please could I speak to him. He was quite charming and said 'stuff and nonsense', of course we could come aboard, that was what they were for... but to hurry up as they'd no idea how long they'd be there.

Mrs Larios and family returned to their lodgings that afternoon to collect some belongings and somehow they persuaded the British Consul to send a car to pick them up and take

London **leads the ships of the First Cruiser Squadron out of Malta, sometime in the late 1930s.**
D.S. Clark Collection.

them down to HMS *London*. The car was 'covered with Union Jacks and communist signs' to protect it from the gunfire of the warring factions.

> *We got down with no trouble at all and on board... and you can guess the kindness of everyone. We stayed on the* London *till 8 p.m.; and were then put on the* Douglas, *a destroyer, and sailed right away. They were shooting on the far side of the quay as we left.*

As the end of July approached the situation ashore went from bad to worse, as Commander Wright noted:

> *29 July. Rumoured that the Communists have been hunting out and shooting the wretched nuns...*

> *30 July. Pretty busy. Number of refugees slightly fewer. Talking to an English lady today who has got onboard with her child and nurse. One of her Spanish brothers-in-law was murdered, the second managed to escape his captives, and her husband made his getaway, from the village where they were holidaying in a cart containing corpses!*

At the beginning of August, the Italian and Swiss governments ordered all their nationals out of Spain. *London* was flooded with desperate people and Commander Wright gave up his cabin to a refugee family, ending up sleeping on the upper deck by X-turret.

London was told on 7 August that she would stay another fifteen days in Barcelona. After returning to Gibraltar she was to head back to Portsmouth for a refit that was due to commence in November, during which she was to receive four more high-angle 4-inch guns. By mid-August it was apparent to Commander Wright that an increasing number of the 'British' refugees were bogus:

> *13 August....The 'British' folk leaving Spain nowadays are not good speakers of English!*

On 22 August he was pleased to note that HMS *Shropshire* had arrived in Barcelona to relieve his ship. Between 23 July and 29 July, 839 refugees were evacuated via HMS *London* and a further 1,000 were sent to safety between 30 July and 21 August.

Meanwhile HMS *London* arrived back at Malta on 24 August but Commander Wright's reunion with his family was to last less than a month – by 16 September, the cruiser was again in Barcelona and was to stay there, lying off a pier, until 13 October. As Commander Wright faithfully recorded in his diary, the slaughter ashore was getting worse:

24 September. Talking to Mr Perks (local resident) who tells me there are 35 bodies of 'bumped off' folk in the local morgue this morning.

On 29 September HMS *London* heard that the Fascists had sunk two Republican destroyers off Gibraltar. Commander Wright recorded how the refugees kept coming:

2 October. Two batches of refuges from Alicante and Palma. The latter included one wretched human whose husband was due to be shot by rebels tomorrow.

3 October. We received warning that, after midnight tonight, the rebel ships are liable to bombard the coast towns from Malaga to Barcelona. 'London' moved to the outer anchorage during first-dog watch. As she did so a refugee bounded his way through the guards on the pier by telling them he was the Pilot! Personally I feel we should have remained inside the harbour.

On 11 October HMS *London* went back into the harbour, putting herself alongside a jetty to take on oil. On the following day she moved back out to the roadstead. Two days later *London* changed over with the *Shropshire* and left Barcelona, heading back to Malta.

Chasing Potato Jones

The *London* was able to witness the other side of the conflict during visits to Palma on the island of Majorca, which was a Fascist stronghold. Italian and German bombers taking part in raids against Republican targets frequently flew over the cruiser from their airfield on the island. Often HMS *London* got to see those same aircraft dumping their bombs on the enemy. On one occasion, while the cruiser was at Valencia to pick up refugees, her sailors looked on as German and Italian aircraft pounded the docks just across the harbour. During a visit to Palma HMS *London* was tasked with hunting down a notorious gunrunner with the quaint nickname of Potato Jones. He was targeted because the British government had decided it didn't want any of its nationals providing arms or even food supplies to either side in the war. Corporal Priestley remembered how the *London* chased in vain after her quarry:

They sounded off action stations... Our task was to catapult the aircraft off. Meanwhile the captain of the ship broadcast what was going to happen. He explained that we'd obtained information that this character was somewhere ahead of us and we'd got to meet him and stop him. So a boarding party was got ready – a Petty Officer with a number of seamen fully armed with rifles, revolvers and cutlasses. It made us RAF chaps giggle a bit but, nevertheless, it was a necessary job and our aircraft was catapulted off to search ahead.

The *London*'s aircraft located a suspicious looking vessel that could well have been Potato Jones on a gunrunning trip. The cruiser caught up after five hours hard steaming, only to discover that it was not the right quarry at all.

In 1937 HMS *London* was called back to Britain for yet another big naval review at Spithead, this time to celebrate the coronation of King George VI.

It was while Corporal Priestley was aboard HMS *London* that the Royal Navy finally wrested control of its aviation from the RAF. Along with thousands of other RAF personnel attached to warships, Corporal Priestley was offered a chance to transfer to the new naval controlled Fleet Air Arm. He declined the offer. Corporal Priestley recalled that the Executive Officer of the *London* wanted to know why none of the RAF personnel serving in her had applied to join the Navy.

HMS *London* arriving at Portsmouth for a short refit. *Strathdee Collection.*

Walrus aircraft from ships of the First Cruiser Squadron flying over Malta in 1938. As the flagship's aircraft, *London*'s Walrus (068) leads, followed by *Devonshire*'s (071), *Shropshire*'s (069) and *Sussex*'s (067). *D.S. Clark Collection.*

Ian Marshall's impressive watercolour 'H.M.S. London, Greenwich 1938'.
Used by kind permission of the Mercers' Company, London.

He sent for me, because at the time I was the senior RAF rating on board. And he asked me why we hadn't applied to transfer to the Navy and I explained that we didn't want to. We wanted to get back to the RAF as soon as possible. He didn't understand that we were posted whether we liked it or not to the Fleet Air Arm. He got a bit of a shock, and I don't think he liked it, because he thought we were being unkind to the Navy. So we just had to carry on, and I think it was approximately nine months later that the ship paid off and we all went back to the RAF anyway.

As the RAF man left, twenty-year-old Writer Harry Williams joined *London*, sailing with her when she returned to the Mediterranean in June 1937:

She was my first ship and I was very impressed – she towered above the jetty like a side of flats. She was good to live in, but, because she was so tall and top heavy, would roll on a piece of damp moss.

Although a member of the team that ran the ship's administration for the Captain, Writer Williams had an action station in the anti-aircraft gunnery control centre.

The London *spent a fair amount of time cruising just outside Spanish territorial waters as the League of Nations was keen to stop people smuggling arms through to the warring factions. This meant us running along a triangle from Palma to Valencia and Barcelona, with*

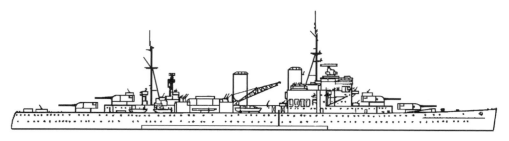

The brutal, boxy lines of the rebuilt HMS *London* **as she looked in early 1941 after work at Chatham Dockyard.** *Dennis Andrews.*

the occasional break in Gibraltar. Half the ship would be at defence watches and I did my share, ready to pass on information to the 4-inch high-angle guns so they could set the fuses of their air burst shells to the right altitude. I remember one time we watched Franco's bombers hitting Valencia, tracking them with the AA guns just in case they thought about having a go at us.
Within a few years *London*'s guns would be firing in anger, but not in the Mediterranean, for her war would be fought in the Atlantic.

Notes

1. R. Ransome Wallis, *Two Red Stripes.*
2. D.K. Brown, *Nelson to Vanguard.*
3. Edward Reynolds, Sound Archives, Imperial War Museum.
4. Edmund Priestley, Sound Archives, Imperial War Museum.
5. Diary of Paymaster Commander N. Wright RN, Imperial War Museum.
6. Papers of Hon. Mrs J. Larios, Imperial War Museum.

Chapter Nine

THE *BISMARCK* HUNT

Sneaking Out of Chatham

With war clouds gathering fast in the late 1930s, and naval construction restrictions being discarded equally rapidly, the decision was taken to reconstruct the County Class cruisers to make them more capable ships.

HMS *London* was the first, paying off into refit at Chatham in March 1939. Her ambitious rebuild proved complex and lasted until February 1941. Among the major modifications were a new waterline armour belt and bridge superstructure, additional 4-inch guns, new hangars and aircraft catapult (the original had been fitted in 1932) and new engines. Her top speed was to remain in excess of thirty-two knots and she would be able to do twenty-four knots using just half her boilers. Having started her life with a displacement of 9,850 tons (standard) and 13,315 tons (deep load), by the end of the refit, HMS *London*'s tonnage was 11,015 tons (standard) and 14,578 tons (deep load).

The severe changes in *London*'s silhouette included the removal of one of her funnels, with the new catapult occupying space between the remaining two. Grouped around the aft funnel were the new 4-inch guns, in twin mountings, with pom-pom anti-aircraft weapons (sixteen 2-pounders) on the hangar roofs and machine guns fitted on top of X and B turrets. Post refit, HMS *London* carried eight 21-inch torpedoes in quadruple mountings on the upper deck and also had depth charges.

The onset of hostilities between Britain and Germany in September 1939 meant the remaining County Class cruisers were too busy to be rebuilt along the same lines as *London*. And, as we will see later, it was just as well, for *London*'s rebuild would not stand up to the strain of war service.

The *London*'s lines were altered dramatically by her major rebuild. She is seen here in her stark Arctic war paint, *circa* **1942.** *R. Ransome Wallis Collection.*

The winter of 1940-41 was a bitterly cold one, particularly if you were living aboard a 'dead' ship in refit. Able Seaman Ken Tamon was a member of the care and maintenance party embarked on HMS *London* in October 1940. He recalled the cruiser was 'a cold, empty shell', in which sailors were kept busy by keen officers who drilled them in some of the essential skills they would need in the war at sea:

> One day the Lieutenant in charge of the duty watch arranged a series of whistle blasts to call us to action. We had to listen carefully to them to interpret what job we were training to do. We thought we were being tested on fire drill, so off we galloped onto the upper deck armed with stirrup pumps and buckets of sand, only to find out we should have been armed to 'Repel Boarders'! The number of whistle blasts had been misunderstood.

Surgeon Lieutenant R. Ransome Wallis was also sent to HMS *London* in the autumn of 1940, after being volunteered by his local medical committee as available for service in the armed forces. Having put himself forward for the Royal Navy, he was ordered to Chatham.

After several weeks of training, Surgeon Lieutenant Ransome Wallis finally joined the *London*. He liked the look of his new ship, despite the carping of others about the County Class.

> Many books about ships of the Royal Navy were written between the world wars and their authors shared in common a vehement dislike of the County Class. They were called 'Tin Clads' and 'expensive white elephants' and it was said that their high freeboard would make them roll violently while their large silhouette would make them easy targets for an enemy.

Surgeon Lieutenant R. Ransome Wallis. *R. Ransome Wallis Collection.*

> Seldom can so-called experts have been so wrong.

London's skipper was Captain R.M. Servaes and one day, shortly before the cruiser left Chatham, he gathered his sailors around him to give them a pep talk. The captain told them there were probably difficult times ahead, but he was confident they would master whatever tasks were thrown at them. Captain Servaes told the crew that *London*'s equivalent was *Prinz Eugen* and '...that a duel between us would prove interesting'.[1] It was not a prospect that provoked much enthusiasm.

Engineering Artificer 5th Class Gordon Bruty had been waiting for a ship at Chatham for five months before he was eventually told to join HMS *London*.

> *Each day we reported to working parties and, in fact, I had seen the* London *many times during our work around the dockyard. My first day aboard was quite a culture shock. I had never seen anything like it – to think that this was our home. During our time waiting for ships we had been given introduction classes that included things like how to sling a hammock, and*

lectures on what shipboard life was going to be like... but it was still a bit daunting. I had never been to sea.

German bombers targeted Chatham frequently and air raids often interrupted the work, delaying *London's* departure. The cruiser was finally commissioned for service with the Home Fleet on 7 February 1941. Her sailors were granted their final shore leave and Ordnance Artificer Graham Bramley brought his wedding forward because he realized he might not be back home for many months.

We were to marry in April 1941, but when I heard the ship was going to be sailing in the early part of the year I put in for some extra leave to get married. I came home on the Friday and got married on the Saturday. That night there was an air raid, but I wasn't going to go spend my wedding night in an air raid shelter, so we didn't take cover. On the Sunday I headed off back to the ship, which left on the Monday. I didn't see my wife for a year.

On 5 March HMS *London* left Chatham. Waiting for her in the familiar waters of the Channel were magnetic mines sown by the Germans and E-boats keen to pounce on her. Ken Tamon recalled that Captain Servaes was not going to let anything get in the way of taking *London* to war:

Graham and Eileen Bramley were married at St Peter's church, Leicester, during weekend leave prior to *London* sailing from Chatham. *G. Bramley Collection.*

He was so eager to get under way and engage the enemy, he ordered all toolboxes and gear belonging to dockyard mateys to be dumped on the jetty, much to the consternation of the workers, some of whom hadn't finished their jobs. Some were still aboard when we sailed.

The cruiser indulged in some simple deception to ensure she got away from Chatham safely, heading for Tilbury instead of going straight out to sea. Graham Bramley recalled:

The Germans used their fifth columnists, spies and the like, plus their scout planes, to keep an eye on Chatham. They seemed to know exactly what was happening. They expected us to come out of the Medway and into the Channel straight away and then head up the east coast of England. But, instead of doing that, the London *hid up the Thames for a day or so to throw them off the scent. The following night we emerged. Lord Haw Haw claimed on the radio they had sunk us but their E-boats had been disappointed.*

Leaving the Thames on 6 March, on her way north, HMS *London* passed a couple of sinking ships which had been unlucky, falling victim to mines or E-boats. The cruiser's first night in the North Sea was a rough one and a number of those onboard, new to seafaring, were rather ill keeping Surgeon Lieutenant Ransome Wallis busy.

Sea sickness was rife and the buckets though strategically placed, were often inadequate.

HMS *London* was headed for Scapa Flow, the Orkneys' anchorage of the Home Fleet, and on the way north managed to fit in some speed trials to confirm she could still do thirty-two knots.

After arriving at Scapa, *London* embarked on an energetic programme of work up exercises. These included sub calibre and full calibre shoots with her main guns at targets and also throw off shoots. The latter entailed her guns having their controls set to throw off

Waldie Willing (right) and a shipmate pose for a photograph aboard *London* **at Scapa Flow. Anchored in the background is the battleship HMS** *Prince of Wales.* W. Willing Collection.

shells by a cable's length (200 yards) astern, even though the weapons were aimed right at the target ship. *London* was soon on the receiving end of this all too realistic gunnery exercise technique, when she acted as a target for the battleship *Rodney's* 16-inch guns. Among other battlewagons the *London* provided a target for, were *King George V, Hood* and *Prince of Wales.*

HMS *London* was a well run 'Pusser's' ship, for there was a higher proportion of regular Royal Navy people in her than many other warships where Royal Naval Reserve and Royal Naval Volunteer Reserve sailors proliferated. 'Discipline was pretty strict and the general high standard made her a smart ship and on the whole, a happy ship.'[2]

There were sixty boy sailors in the ship, many of them from borstals and tough communities. Some of the boys did very well during their time in the cruiser, but others were hard to handle. The more serious offences committed by these tough nuts were punished with flogging:

> *...or more accurately so many strokes with a thick and heavy cane. This was quite a performance and was enjoyed by nobody...On the whole I doubt if these floggings did any good.*[3]

On 2 April, the *London* left Scapa to rendezvous with the carrier HMS *Argus*, which was being used to ferry fighter planes to the Mediterranean. The cruiser escorted *Argus* to a point 300 miles west of Lisbon where she was turned over to the care of HMS *Sheffield.*

Search and Destroy

HMS *London* spent some time with the battleship *King George V* keeping watch on Brest to ensure the German battlecruisers *Scharnhorst* and *Gneisenau* remained bottled up. The two German heavy ships had taken refuge at the French naval port in February 1941 and would be there for another year. Relieved of this duty, *London's* next mission was escorting slow convoys out of Sierra Leone. This was difficult for the ship to deal with, as the merchant vessels could do little more than six knots and she tended to wallow at such a slow speed.

Slow convoy to Sierra Leone, as pictured from HMS *London*. *R. Ransome Wallis Collection.*

> *Generally we sat in the middle of the convoy with a couple of destroyers and an armed trawler discretely* (sic) *arranged on the flanks. Flying fish and dolphins were frequent and apart from a few submarine scares the whole thing had a bit of a holiday atmosphere.*[4]

To relieve the tedium, and provide a means of boosting the confidence of the merchant ships, HMS *London* ran up and down the convoy lines at twenty-two knots, then took a circuitous route, before going back to the centre.

On 12 May, at a UK port, *London* embarked soldiers who were being taken south to Gibraltar. The generous internal proportions of the County Class cruisers made them ideal troop carriers and one of the young Army officers aboard was John Cunningham:

> *I had just been commissioned into the Black Watch and was posted to join the 4th Battalion in Gibraltar. While aboard* London *we slept in the officers' quarters and, as far as I can recollect, the food was good and compared well with our Officer's Mess fare. We didn't have a lot to do, as it was a fairly idle time. We played cards a lot, mostly Brag. Apart from the tug-o-war and deck quoits I seem to remember doing quite a bit of jogging 'round the deck. I didn't find life at sea too bad because I am a reasonably good sailor, so I wasn't seasick. Some gun practice, with the 8-inch weapons, took place on one occasion and we were issued with cotton wool to protect our ears. After some time in Gib I suffered severe earache and it was discovered that a piece of the cotton wool had impacted against my eardrum.*

After offloading her 600 Army passengers at Gibraltar, *London* stayed in port for a few days, as she was due to return to Britain carrying soldiers going home on leave. Together with the cruiser HMS *Edinburgh*, she was also to provide escort for a UK-bound convoy that included the liner SS *Arundel Castle*, carrying evacuees from Gibraltar. Shortly before the cruiser left, Italian aircraft tried to bomb the Rock, but hit a nearby Spanish town by mistake, killing a number of people. *London*'s AA guns went to action stations but the planes were too far away to shoot down.

Sailing north from Gibraltar, on 23 May, *London* encountered a Vichy French ship that she stopped and searched. The skipper of the French vessel bore no grudges for he,

> *...presented us with a hogshead of French* vin ordinaire. *The French usually dilute this before drinking it which is a wise precaution as we found out later. Partaking of the neat brew made us quarrelsome and the wine became known as 'Red Infuriator'.*[5]

To occupy themselves during the voyage south to Gibraltar the young soldiers of the Black Watch challenged *London*'s officers to a tug of war contest. John Cunningham is one of those pulling (second from the front). *J. Cunningham Collection.*

The view forward over *London*'s guns in the South Atlantic, summer 1941. *W. Willing Collection.*

Making the hangovers of *London*'s sailors worse the following day, was the blackest news of the war so far for the Royal Navy – the German battleship *Bismarck* and heavy cruiser *Prinz Eugen* had broken through the Denmark Strait, destroying the battlecruiser *Hood* and damaging the new battleship *Prince of Wales*. While the *Bismarck*'s 15-inch guns finished the *Hood* off, the *Prinz Eugen* – *London*'s counterpart – had managed to inflict considerable damage on the vulnerable old battlecruiser's upperworks with her 8-inch guns. Only three sailors out of *Hood*'s 1,418-strong crew survived. Like everyone else on *London*, Surgeon Lieutenant Ransome Wallis was afflicted with great sadness:

> *...most people having had friends in her. I had known her First Lieutenant as a boy. I also recollected sadly a very young midshipman who had spent a couple of days in* London's *Sick Bay. He had recently been sunk in another ship and had only survived after a bit of an ordeal. He said to me rather bravely 'But I shall be all right now sir, I am going to the* Hood'.

Damaged by shells from the 14-inch guns of *Prince of Wales*, the *Bismarck* should have immediately turned for home. Instead both German warships pressed on into the Atlantic, where supply vessels were waiting with the provisions they needed for an extended campaign against British merchant shipping. The German surface raiders hoped to be aided in their fight by converging with seven U-boats using the same supply ships. The County Class cruisers *Suffolk* and *Norfolk* were providing steady reports for Home Fleet Commander-in-Chief Admiral John Tovey, as they were shadowing *Bismarck* and *Prinz Eugen*.

> *The British Admiralty meanwhile, fast recovering from the shock of the* Hood, *were also making dispositions, and not for the first but almost the last time in the Royal Navy's long and brilliant history were exercising the strength that global sea-power gave it.*[6]

British ships were vectored in from all points of the compass. The old battleships *Revenge*, and *Ramillies* from Canadian waters, the *Rodney* heading out from the Clyde with her destroyers and cruisers, while Force H – including the battlecruiser *Renown* and carrier *Ark Royal* – left the Mediterranean. Most important of all, the battleship *King George V*, the carrier *Victorious* and battlecruiser *Repulse* plus five cruisers, were within striking range. The *London* was also called away from her convoy, with orders to join the chase. Shortly before 5 p.m. on 24 May she received a signal, which said:

> *Part company with* Arundel Castle *and destroyers. Order them to proceed in execution of previous orders.* London *proceed at economical speed...your movements should be adjusted to close enemy and you should prepare to take over shadowing duties.*[7]

Engineering Artificer Bruty noticed the change in pace:

> *Down in the engine room of course we didn't know what was going on. We were aware that something was up but not the exact nature of it. The machinery was working at high speed and you wouldn't do that for nothing. I think that we were told eventually that we were taking part in the hunt for the* Bismarck.

After *Ark Royal*'s Swordfish inflicted damage that slowed her down, the *Bismarck* was cornered on 27 May. *Rodney* and *King George V*'s big guns did the main damage, while torpedoes from the cruiser *Dorsetshire* completed her destruction. *Prinz Eugen* had parted company with *Bismarck* on 24 May and, after a week prowling the Atlantic, she had headed for Brest.

Cracking Enigma

This left the supply ships to be hunted down and destroyed, with HMS *London* one of several British cruisers ordered to scour the central and southern Atlantic.

At 9.19 a.m. on 25 May the *London* had received further instructions by signal from the Admiralty, diverting her away from the *Bismarck* hunt. A general signal said: '*London*... proceeding to search for enemy tanker.'

Aside from tidying up loose ends from the *Bismarck* chase, it was important that the supply vessels should be eliminated to hamper the activities of U-boats and also ensure the heavy ships trapped at Brest would have no source of supply should they break out into the Atlantic. Great excitement was caused aboard *London* when the cruiser's radar made a promising looking contact:

> *The Walrus was flown off to investigate this blip, which turned out to be a U-boat giving its crew a swim. He dived before the aircraft could attack successfully, but the* London's *Executive Officer – 'Trunky' Edden – said over the tannoy that the Germans had not kept a lookout 'with usual German inefficiency'! There were loud groans in the mess-decks, especially from people in our crew who were survivors of the Armed Merchant Cruiser* Jervis Bay, *which had been sunk with so-called 'German inefficiency' the previous November.*[8]

On 31 May, HMS *London* called in at Bathurst, Gambia, to land some of the troops embarked at Gibraltar. They were grateful to escape the confines of the cruiser, but not exactly pleased at being further away from Britain than when they had started. Their chances of spending their leave with their families looked rather slim. HMS *London* sailed again that afternoon, heading for a rendezvous on 3 June with the destroyer HMS *Brilliant*, which would be her hunting partner. The following day at 07° 35′ N – 31° 29′ W, the two British warships found their first German supply ship, the 17,000 tons *Esso Hamburg*, which had, in fact, refuelled the *Prinz Eugen* on the morning of 28 May. The oil transfer had been curtailed when the German cruiser's lookouts spotted smoke on the horizon.[9]

A telltale column of smoke rises from the sinking *Esso Hamburg*. *R. Ransome Wallis Collection.*

Six days later, Ordnance Artificer Bramley helped man the *London*'s 8-inch guns.

> *During this time I was in Y turret. You couldn't see anything, just the mechanism working and the sound of the gun going off. My job was to oversee and maintain the hydraulics. When we found the* Esso Hamburg *she started heading away from us and I remember a Petty Officer in Y turret saying 'for tuppence I'd put a shell through them buggers'. But instead, the* London *used her two-pounder pom-pom to put some shots across her bows. It put the fear of God in them. When the man who had fired the two pounder got back home he found that his family had been killed in an air raid. I am sure if he had known he would have made certain his warning shots hit the German ship.*

Having secured from action stations, with the exception of the gun crews and some damage control teams, many of *London*'s sailors made their way to the upper deck to watch what happened next. Surgeon Lieutenant Ransome Wallis was among those looking on:

> London *was steaming about 22 knots in a large circle about a big tanker, which was obviously in its death throes, listing badly and down by the bows. Her crew in two large lifeboats were pulling towards us and we stopped for a short time to pick them up before continuing to circle the stricken ship, which now suddenly capsized and floated bottom upwards. Her underside was painted a bright red.*

Boy Seaman Waldie Willing witnessed the scene while crewing one of *London*'s 4-inch guns:

> *At first the Germans looked like they were striking out for the Fatherland. One of our officers got this loud hailer and used it to persuade them to come alongside, which they did and came up the scrambling nets. We were keen to get them aboard as soon as possible, as we were stopped dead in the ocean, a sitting duck for any U-boat.*

The prisoners of war picked up by the *London* were German Navy sailors, with a sprinkling of merchant seamen. Able Seaman Tamon heard one of the captives expressing gratitude at being 'liberated':

> *He was a Polish messboy called Karel, who was interrogated by two of our Polish Midshipmen. It turned out he had been forcibly conscripted by the Germans in Gdansk and went to sea rather than serve in the army.*

One of the *London*'s sailors asked a German officer how he thought the war was going for the Reich. With great pomposity, the German replied: 'We shall win on the land, in the air and on the sea.' But, on glancing over at his sinking ship, he corrected his statement: 'No, we shall win on the land and in the air.'[10]

But the *Esso Hamburg* was being very stubborn, so HMS *Brilliant* fired a torpedo into the supply ship, which failed to go off. Next, she poured fire from her 4.7-inch guns into the German vessel, which caught fire and sank. Unfortunately the huge pillar of smoke from burning oil curling up into the sky was a perfect warning sign for any other German vessels lurking in that stretch of ocean.

Back aboard *London*, the German officers had been given the gunroom as their accommodation. It had been carefully prepared before they moved in:

> *...a couple of rather crudely hidden microphones and a continuous listening watch by German speaking ratings was initiated. However the only interesting thing that came to light was that they thought we were a battleship and that there was another German ship near at hand.*[11]

German speaking members of *London*'s crew – including the Polish midshipmen – were also put in the gunroom, to pose as prisoners from another supply ship, to see what they might pick up. More powerful than eavesdropping, or effective than sailors posing as Germans, was the intelligence being provided by Bletchley Park, the renowned, but very secret, code-breaking centre in the UK, where Germany's Enigma encryption had just been cracked.

On 9 May 1941 to the south of Iceland, *U-110* had been forced to the surface and her crew abandoned ship, believing they had successfully set scuttling charges. However, the submarine failed to go down and a boarding party from the British destroyer HMS *Bulldog* managed to recover an Enigma machine, rotor settings, charts and code books. This amazing episode was kept secret until the early 1970s, when it was finally admitted that Britain had been able to read encrypted German signals throughout the rest of the war, gaining a decisive advantage for the Allies. Although a Luftwaffe Enigma machine had been passed to Britain by the Poles early in the war,[12] *Bulldog*'s boarding of *U-110* was a key moment. It yielded material that enabled penetration of signals restricted to German naval officers and access to a proportion of sighting reports sent back to Germany by U-Boats in the Atlantic. While radio direction finding and British spies in Germany, Spain and Portugal provided essential information in the hunt for the supply ships, cracking Enigma helped fix the planned U-boat rendezvous points. The cruiser HMS *Neptune* cornered the *Gonzenheim* on the same day as *London* dealt with the *Esso Hamburg*, while the cruisers *Kenya* and *Aurora* sank the *Belchen* the previous day.

On 5 June HMS *London* found the supply ship, *Egerland*, which was flying the Panamanian flag. The British cruiser's crew believed they had found their latest victim via their ship's radar. For, like everyone else, they were to remain ignorant of Bletchley Park's work for thirty years. Surgeon Lieutenant Ransome Wallis witnessed the drama of the *Egerland*'s interception:

> *In an effort to prevent her from scuttling herself* London *opened fire on her at a very long range and it appears that our first shot went through her bridge and set her on fire. At least the gunnery officer always stoutly maintained this was so.*

The German vessel's crew set scuttling charges and abandoned ship, with several more boatloads of Germans taken aboard the already crowded *London*. Many of the new POWs were U-boat crews who had been waiting for submarines to turn up. The new intake contained

Prisoners from the *Egerland* on *London*'s quarterdeck. *R. Ransome Wallis Collection.*

some arrogant individuals who appeared to epitomize the Nazis' Aryan ideal – they were tall, blond, tanned and very fit. But, as Surgeon Lieutenant Ransome Wallis discovered on examining them, they nearly all had bad teeth. When they became afflicted with severe food poisoning, the Germans thought they were being finished off to ease the overcrowding. But the British were equally ill, as Surgeon Lieutenant Ransome Wallis wryly noted:

The cause was found to be a ham which had been boiled and served to most of the ship's company. One of the hams had a diseased bone and this had contaminated all the rest. For a time it was difficult to find enough men to run the ship and all sorts of bizarre possibilities suggested themselves.

Sickbay berths full of British servicemen provided ample proof that it wasn't an attempt to assassinate the POWs.

The senior German officer was shown the rows of sufferers in and outside the sick bay, in order to convince him that there had been no attempt to poison his countrymen.[13]

Blessed with enough able bodied sailors to keep her underway, HMS *London* headed to Freetown in Sierra Leone to offload her passengers and for part of the way she helped escort a convoy heading in the same direction. One of the ships in this convoy was an old liner called the MV *Adda*, which was carrying nurses 'with whom we exchanged many friendly waves...'[14] *London* left the convoy on 7 June and went ahead, with the *Adda* close behind, to Freetown.

...but darkness had fallen before she got to the entrance of the estuary. For some reason her Captain decided to hang about outside until daylight and this proved to be a fatal decision as a lurking U-boat got her with a torpedo early on June 8th and there were only a few survivors, most of the nurses being drowned.[15]

During a brief stay in port, the *London* offloaded the remainder of her British troops onto HMS *Norfolk*, which was heading back to the UK, while the Germans were sent to prison camps. Back at sea, HMS *London* resumed her search for German supply vessels. On 12 June the *Friedrich Breme* had been sunk by the cruiser HMS *Sheffield* while, on 15 June, the *Lothringen* had been located and destroyed by planes from the carrier *Eagle* assisted by the cruiser *Dunedin*. Six days later, close to the equator, HMS *London* found the *Babitonga*,

...which had left Santos, Brazil, on 24th April for Brest. This ship had been working with a German merchant raider and was disguised as the Dutch Japara. On being intercepted, she scuttled herself, about 930 miles south-west of Freetown.[16]

HMS *London* had been sailing for a few days in company with the destroyer HMS *Highlander* but she was on her own, and not anticipating any action, when the *Babitonga* was discovered.

Babitonga **going down by the stern.** *R. Ransome Wallis Collection.*

The *Babitonga's* **scuttling charges explode.** *R. Ransome Wallis Collection.*

We were conducting our Crossing the Line Ceremony, which was unceremoniously terminated by the sighting of this German tanker flying the Dutch flag.[17]

As it appeared the small German merchant ship posed no threat, *London's* off watch sailors were allowed to come up onto the upper deck to watch her being sunk.

We approached cautiously and suddenly there was a considerable explosion as her scuttling charges went off...She quickly stood on end and went down just as her survivors were being taken aboard.[18]

Engineering Artificer Bruty was amazed to find a former countryman among the *Babitonga's* crew.

This chap came up the scrambling net and said something to the effect of 'good morning gentlemen' in perfect English. Turned out he was a Briton who had become a naturalized German.

As they came aboard the British warship, the POWs were intrigued by the strange manner of dress adopted by their hosts.

Babitonga **survivors coming alongside HMS** *London.* *R. Ransome Wallis Collection.*

> *During the Crossing the Line Ceremony you dress up in exotic and strange rig as members of King Neptune's Court. We had dashed to action stations without changing and when the* Babitonga's *captured crew came aboard they must have thought they had been run down and captured by a ship of cannibals.*[19]

Not deterred by pagan costumes, one of the *Babitonga's* crew decided on a show of defiance as he picked himself up on *London's* upper deck:

> *He did the Hitler salute as he came aboard, so this Royal Marine let his rifle accidentally slip and bang the German on his head. There were no more Hitler salutes.*[20]

On the whole, the new batch of prisoners taken aboard *London* appeared to be amiable merchant seamen. In reality they were German Navy sailors and the *Babitonga* had been far from harmless, as she was an armed merchant supply ship working with an ocean raider called the *Atlantis*.

Boy Seaman Willing didn't think much of the Germans he was sharing space with:

> *We ended up with 138 POWs and they took over the boys mess deck and some of them were a bit arrogant. We didn't bother with them, as they were looked after by the marines. The Germans wanted this and that, such as cigarettes, and mostly lounged on the upper deck in the sun.*

No more German supply ships were located and so *London* returned to Freetown to offload her prisoners. By this time some of *London's* sailors had gone down with malaria and venereal disease, the latter contracted from consorting with prostitutes in Sierra Leone. Surgeon Lieutenant Ransome Wallis had cautioned them against this, but the sailors in question had stubbornly ignored his advice:

> *The local dives were pretty terrible but it was hard to convince some of the ratings that they were in any danger.*

On 6 July, HMS *London* left Freetown and four days later came across another Vichy French merchant ship that was boarded and searched. Two weeks later, during a visit to Gibraltar, *London* was suddenly called back to home waters. Steaming at top speed for several days, the cruiser was needed to help contain another German surface raider believed to be readying herself for a breakout into the Atlantic.

Notes

1. R. Ransome Wallis, *Two Red Stripes.*
2. ibid.
3. ibid.
4. ibid.
5. ibid.
6. Ludovic Kennedy, *Pursuit.*
7. Jack Broome, *Make Another Signal.*
8. Ken Tamon, in a letter to the author.
9. Fritz-Otto Busch, *Prinz Eugen.*
10. R. Ransome Wallis, op. cit.
11. ibid.
12. John F. White, *U-Boat Tankers 1941-45.*
13. Ken Tamon, in a letter to the author.
14. R. Ransome Wallis, op. cit.
15. ibid.
16. HMS *London, Summary of Service,* Royal Navy Historical Branch.
17. Ken Tamon, in a letter to the author.
18. R. Ransome Wallis, op. cit.
19. Graham Bramley, in an interview with the author.
20. ibid.

Chapter Ten

THE KREMLIN DELEGATION

Friendly Invaders

The high seas raider alert proved to be a false alarm and, after loitering off Iceland for a short while, *London* came down to Scapa Flow, arriving on 23 July. The cruiser only spent a week in the Orkneys before sailing for Greenock, in the mouth of the Clyde on Scotland's west coast, where her crew were awarded some much needed shore leave. Following this, *London* was assigned as escort to a convoy that contained a number of large troop ships bound for North Africa via the Cape.

On 10 August, just south of the Azores, *London* was relieved of escort duty by another cruiser and headed back to the UK, arriving at Scapa Flow eight days later. After another visit to Greenock, *London* was ordered north again to Iceland, arriving at the Hvalfjord on 2 August. Hvalfjord, about fifty miles from Reykjavik, was a major Allied naval base with another fleet anchorage at Seydisfjord. Iceland's importance as a staging post for convoys could not be underestimated. Possession of a secure base of operations for aircraft and warships in the Arctic was vital in reducing the protection gaps in vast stretches of ocean.

Iceland was a Danish protectorate and, after the Germans invaded Denmark and Norway in April 1940, the British occupied the island and used it as a base. The Danish territories of Greenland and the Faeroe islands were also made available.

The Icelanders were initially deeply unhappy about the foreign invasion, but they soon learned to put up with their friendly occupiers. The Americans took over administration of Iceland in July 1941, building substantial military bases encompassing all the comforts of Hometown, USA. Although America was not officially in the war at that point, it did undertake to guarantee the protection of shipping sailing between the USA and Iceland; a clever way for President Roosevelt to help fight the Nazis without actually declaring war. Roosevelt recognized that the

HMS *London*'s, mission to northern Russia was agreed in August 1941, during a meeting between Churchill and Roosevelt on the USS *Augusta*.
US Naval History and Heritage Command.

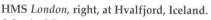

HMS *London*, right, at Hvalfjord, Iceland.
G. Bramley Collection.

future of democracy was at stake and that America should be committed to helping Britain defeat German totalitarianism. Getting in the way were very powerful isolationist, even pro-German, lobbies in the USA that wanted to keep America out of the conflict. In fact, Roosevelt had conned the country, winning the 1940 presidential election by promising to keep America out of the so-called European war. Once re-elected, Roosevelt used stealth to support Britain, via Lend-Lease legislation that provided the full resources of US industry to supply armaments, in exchange for British military bases, including Iceland. The month after the Americans moved into Iceland, the Atlantic Charter saw the USA and UK signing up to shared ideals, including fighting against world tyranny and preserving economic and democratic freedoms.

In late September 1941, HMS *London*, now back at Scapa, went on a journey that would add a new dimension to war in the North Atlantic. She received orders to embark on an Anglo-US diplomatic mission that was to be taken to Archangel in the far north of Russia. The final destination of this important group was the Kremlin in Moscow, to draw up an agreement with Soviet leader Joseph Stalin for cooperation against Germany. The prospect of joining forces with the Russians did not fill *London*'s crew with joy.

To a good many people the idea of any sort of alliance with the Soviet Union was repugnant. But whatever one felt about the Bolsheviks, it must have been plain that, given proper support, here at last was a vast and powerful nation which could stand up to the German war machine.[1]

The Soviet Union, which had eagerly helped the Germans carve up Poland in 1939, was invaded by Hitler's legions in June 1941 and by September things looked pretty desperate. Millions of Russian soldiers had been killed or taken prisoner and German divisions were within striking distance of Moscow itself. Repellent though Stalin's regime might be, both America and Britain realized that helping the Soviets persevere through their most dangerous moment would ensure that Germany faced its mightiest foe – the Russian winter. In the immediate aftermath of the German invasion, Churchill, ever the canny war leader, had committed his country to supplying whatever the Russians needed. The USA decided it

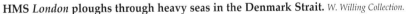

HMS *London* ploughs through heavy seas in the Denmark Strait. *W. Willing Collection.*

would also supply the Russians. Both Churchill and Roosevelt felt that a formal agreement should be drawn up during a joint meeting with the Russians, which was called the Three Power Conference.

The Anglo-American mission to Moscow left Scapa Flow aboard HMS *London* on 22 September. The cruiser carried Lord Beaverbrook, the Canadian newspaper tycoon who Churchill had made his supply minister, as Britain's chief negotiator, with two British Army generals, Air Ministry officials and secretaries making up the rest of the UK party. The Americans were led by business tycoon and diplomat Averell Harriman, who had been sent to Britain by President Roosevelt to oversee administration of the Lend-Lease agreement.

The weather during the voyage turned out to be calm and the Germans failed to find the *London*, which had sailed without escorts. She deposited her precious cargo at Archangel on 27 September, having sailed through the White Sea and thirty miles up the broad River Dvina. From Archangel airport the delegation flew directly to Moscow. Meanwhile, on the Dvina, *London*'s curious officers tucked into some caviar which had been presented to them as a welcome gift. It made them feel quite nauseous.

On 28 September, *London* left Archangel, which was not a good place to hang around as it was being bombed regularly. To kill time, until the diplomats were ready to be picked up for the journey home, the cruiser accompanied a UK-bound convoy of half a dozen ships. A month earlier they had made the first Arctic convoy journey of the war – codenamed Operation Dervish. Now they were carrying timber back to the UK. German aircraft flying from airfields in Finland could be heard hunting for the convoy throughout the time *London* was with it. Newly promoted Petty Officer Ordnance Artificer Graham Bramley recalled that low cloud ceiling and fog provided perfect protection:

> On our radar we could see these German planes approaching but we entered the fog, which was lucky for us. As they flew over we trained the 4-inch AA guns just in case. Unfortunately someone fired one. We thought we were in for it then, that the Germans would see the shell burst and then get a fix on us and start attacking. But the fog was too thick by half. We were sitting ducks escorting those cargo ships, which I think were carrying timber. The London was in the middle of them and doing only five knots, so if it hadn't been for the fog the German planes would have caught us cold.

Handing over convoy escort duties to HMS *Shropshire* on 2 October, the *London* turned around and headed back to Archangel. Seaman Gunner Bob Houghton, who would later serve on HMS *London*, was a sailor aboard the Archangel-based mine sweeper, HMS *Harrier*. He looked on with some amusement as the mission members went across to the cruiser in a most undignified fashion:

> The London was anchored in the White Sea and unfortunately there was somewhat of a gale blowing with heavy seas and high winds. Our skipper had a terrible time trying to get alongside. Eventually, by making use of one of London's forecastle mooring wires we finally secured 'port side to', sustaining some damage to our ship's side and 5-inch armament. What made it really difficult was the high freeboard of London, and ours being several feet below, plus big rollers streaming down the ship's side. One moment we were level, the next we were a fair old drop down from London's upper deck. How on earth were we going to get them across the frightening drop? It was resolved as follows: On the deck above our 'sweep dish aft' we had a flat top locker used for storing our spuds...when we could get any!...So a gangway was rigged from the top of that to the upper deck of London. With a heaving line around each person's

waist, every time the sides were more or less level, these very alarmed VIPs scurried across to safety! I'll bet they never forgot that little experience.

Over a restorative cup of tea in the wardroom Surgeon Lieutenant Ransome Wallis asked the Mission members how they had got on:

They seemed glad to be back. I gathered that the military liaison idea had been a complete flop as the Russians refused to talk about military matters, but the rest was satisfactory, if such a term can be used when one is giving away vast quantities of armaments which could be ill spared. It was also agreed that Great Britain and America would help in the transportation of these armaments to Russia.

The Spy Who Came in From the Cold

Some extra people had tacked themselves on to the mission, including a journalist and Corporal James Allan of the Royal Military Police. The soldier was sent down to the sickbay for a check up and Surgeon Lieutenant Ransome Wallis heard an incredible tale. Allegedly captured by the Germans in the summer of 1940 at Calais, Corporal Allan claimed that he had escaped from a prison camp and headed across Poland to Russia.

...the Russians suspected him of being a spy and locked him up. They knocked him about and generally ill-treated him for a time but eventually released him from Lubianka Prison. He made his way to the British Embassy and was passed on to accompany the Mission back home.

Graham Bramley recalled that Lord Beaverbrook later asked for the Corporal to be brought to his cabin so they could have a chat.

The Corporal's clothes had been in tatters so they took him down to the marines who gave him some kit to wear and he had a bath and shave too. So, when this man dressed in a marine uniform arrives, Beaverbrook sits him down and asks him if he would like a drink. The chap says 'a whisky would be nice!'. Beaverbrook asks him if he would like a smoke and offers him a cigar, which he takes and lights up. Beaverbrook starts chatting to him, asking him to tell him more about how he was treated in Russia. The man replies, 'I didn't go ashore'. Beaverbrook looks puzzled and asks him what he means by that. Beaverbrook's cabin guest explains that the London *didn't stay alongside. Beaverbrook still looks puzzled. The man gets up and explains that he is a marine orderly and he has come to close the cabin's scuttles for darken ship. So you see Beaverbrook has mistaken this orderly for the soldier rescued from Russia. He sees the funny side of it though and suggests the marine goes and fetches the soldier and brings him to the cabin. I heard all about it because the marine came back below chomping this cigar and we asked him where he got it.[2]*

Some members of the Mission were highly suspicious, suspecting Corporal Allan of being Stalin's spy. Surgeon Lieutenant Ransome Wallis witnessed attempts to catch him out:

Although he was very well looked after in the Sick Bay, he was surreptitiously watched by a guard and was questioned by the Royal Marine officer in charge of security. In fact when he got back to Scapa an attempt was made to make him give himself away by pretending that we were back in Russia and a Royal Marine Lieutenant who he had not seen, wearing a rain coat and a cap with a red band, came in and stared at him. Allan however was blissfully unconscious of all this. His story was absolutely true and he was later awarded a DCM but without a citation so as not to offend our new ally.

While they might have endured a frightening experience in rough weather getting back aboard *London* in the White Sea, the members of the Anglo-American mission to Moscow were lucky with the weather. It was calm on the way there and not too bad on the way back. The ease of the journey was something that might have lulled the Allied delegates into agreeing too readily to supply Russia via the route around northern Norway to Archangel and Murmansk. There had been three possible routes on the table; across the Pacific to Vladivostock and then by rail across Siberia, via the Gulf to Basra and then up through Iraq and Iran or via Murmansk and Archangel.

In reality, the Gulf-Basra route was used to supply the Soviets with much of their supplies, but the most important route politically was the north Russia route. The Russians had pressed heavily for it because they needed a second front opened against the Germans in Europe.

'This was the route that Stalin demanded should be used as the main supply route and the mission accepted this without much hesitation.' Surgeon Lieutenant Ransome Wallis later lamented. 'One wonders to what extent they had been influenced by their peaceful and enjoyable trip in HMS *London* where no effort had been spared to make them comfortable.

Now had HMS London *been shadowed by a Focke-Wulf aircraft, attacked by U-boats at frequent intervals and had to fight off hordes of high-level bombers, dive bombers and torpedo-carrying aircraft for the last five days, one wonders if the Mission would have accepted Stalin's demands so readily.*

White Hell

For *London* many months in the Arctic lay ahead, with the cruiser pushed to the limits of her construction and her sailors to the edge of human endurance. The Denmark Strait, between Iceland and Greenland, was a key patrol area where cruisers based in Iceland, including HMS *London*, would loiter, waiting for attempts by German capital ships to break out into the Atlantic. As happened with the *Bismarck* and *Prinz Eugen*, the cruisers would be expected to tail the enemy, fixing their position for air strikes by carrier planes and subsequent destruction by Royal Navy battleships. In terms of weather, Patrol White was possibly the worst duty of the entire war for British cruisers.

The Denmark Strait is only about 100 miles across in the summer and in the Arctic winter, when the sea freezes, it narrows to about 50 miles wide. It's the point where the Atlantic flows into the Barents Sea like a water spout full of violent seas. A ship doesn't just roll...it twists...up there a ship really labours...it creaks and groans all the time. Riveted ships have enough give in them to take it, but welded vessels, like some of the American warships, can't take it and break. London *had areas on her that were welded and they were a problem.*[3]

It was the major reconstruction that made *London* brittle and prone to cracking. The armour belt and other additions had added so much weight and were so inflexible that,

...reports were received of leaking rivets and cracks in the upper deck, particularly around the boiler room uptakes.[4]

During a refit at Palmers on the Tyne, which lasted from November 1941 until January 1942,

...an extra 63 tons of strengthening was added to the upper deck. These repairs were ill-conceived as strengthening the upper deck raised the neutral axis increasing the stress on the bottom. Sure enough, the bottom began to leak, contaminating the feed water... .[5]

During the refit, although there were some bad air raids, *London*'s crew enjoyed a restful period. But, while *London* was in refit the rest of the Royal Navy was enduring some of the worst disasters of the war so far. The battleship HMS *Barham* blew up after being torpedoed in

Even the King George V Class battleship HMS *Duke of York* could find the weather in the Arctic hard to handle. *Strathdee Collection.*

the Mediterranean in late November and in mid-December the battleship *Prince of Wales* and the battlecruiser *Repulse* were sunk in a foolish bid to prevent the invasion of Malaya. But, at least the Japanese attack on Pearl Harbor on 7 December had brought America fully into the war and it was likely that the US Navy would contribute units to the fight in the Arctic.

Surgeon Lieutenant Ransome Wallis observed that *London*'s sailors dreamed of a release from the hell of Patrol White:

> As our refit neared its completion the usual crop of buzzes about our future movements occurred. There was a good deal of wishful thinking and the Indian Ocean, South Atlantic, Mediterranean and the West Indies were all hopefully put forward. However, there was an ominous issue of sheep skin coats and dogskin caps. My cap apparently had once belonged to a fox terrier.

On leaving refit the *London* duly headed back to Scapa, much to the disappointment of the crew, and then, after working up in early February, to their horror, she sailed for Iceland and straight out on a dreaded Patrol White '...for about two weeks in vile weather and nearly total darkness'.[6]

The next few months were a living hell of Patrol White with the occasional respite of convoy escort work. The fury of the Arctic winter in those turbulent waters between Iceland and Greenland almost defies the imagination, but Nicholas Monsarrat, a veteran of the Second World War's Arctic convoys, captured it in all its terrifying majesty in his novel, *The Cruel Sea*:

> It was more than a full gale at sea, it was nearer to a great roaring battlefield, with ships blowing across it like scraps of newspaper... a ship was scarcely a ship, trapped and hounded in this howling wilderness.

Monsarrat's experience of the Arctic was gained in a frigate less than a third of *London*'s size, and therefore without the benefit of her high freeboard. But the big cruiser was still powerless in the face of the same savage forces of nature so memorably described by Monsarrat:

> Huge waves, a mile from crest to crest, roared down upon the pigmies that were to be their prey; sometimes the entire surface of the water would be blown bodily away, and any ship that stood in the path of the onslaught shook and staggered as tons of green sea smote her upper deck and raced in a torrent down her whole length.[7]

Charles Cox was one of *London*'s cooks and a keen amateur boxer who often won boxing competitions for his ship. During the Arctic gales the galley in which he worked was as dangerous as any boxing ring, and was even rigged up like one:

> *We had ropes along the walls and across the galley, to give us something to grab onto while we were cooking. It could be a dangerous place. You would take these boiling hot pots and sizzling trays and put them on the deck and the ship would move violently, sending them flying across the galley floor and fat would spill out of pans, hit the hot plates and catch fire... but despite all of that we never failed to cook a hot meal.*

Norman Brigden remembered that *London*'s bakery was a source of considerable comfort to the crew even if they didn't fancy anything else the cooks produced.

> *To be honest, I did not think much of the food we had in* London*, but the white bread they baked was the best I've ever tasted before or since.*

On Sunday the sole cook on duty in the bakery made a special effort – turning out 1,000 rolls on top of his usual quota of loaves. 'Those fresh rolls were vital to ship's morale,' recalled Charles Cox.

In the milder weather members of the ship's rat population, which normally stuck to the spud lockers and other small, warm places during the worst Arctic weather, came out to watch the cook at work with his dough.

Injury was not just a constant hazard for the cooks. Anybody could be flung against bulkheads while making their way about the ship and there was '...seasickness that could exhaust a man beyond the wish to live'.[8] According to Charles Cox, the cooks in *London* were immune to it:

> *One thing that has always amazed me was that not one of the 15 cooks in* London *was ever sea sick, which is incredible when you consider all the smells and conditions in the main galley.*

The ocean was merciless with a warship's structural weak spots, as Surgeon Lieutenant Ransome Wallis discovered.

Ice coats HMS *London*'s forecastle. *W. Willing Collection.*

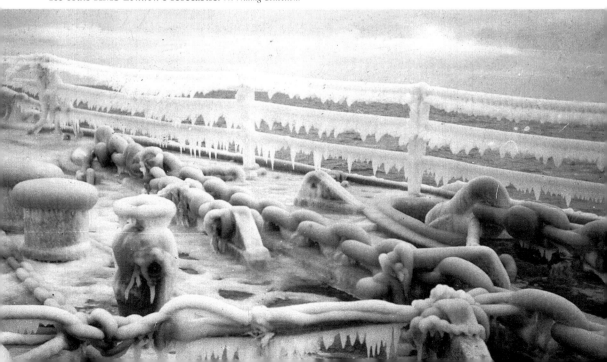

My cabin scuttle on the main deck was 24 feet above the waterline but on one occasion a wave smashed the inch-thick glass and the heavy steel deadlight behind it and put an enormous amount of water into my cabin.

The sea would reach up over *London's* high freeboard to snatch whatever victims it could find:

The Chief Bosun's Mate ordered a Leading Hand into the seaboat to prepare to turn it inboard. 'What me, Chief?' he asked a little nervously. Just then she gave an enormous roll, sea up to guardrail, and, as we righted, the seaboat was gone![9]

The *London's* upper decks could swiftly be covered in treacherous ice, the spray having frozen as it hit the ship. This had to be removed, as the extra weight could render the weapons inoperable or even make the ship capsize. Petty Officer Bramley was one of those who had to go out and do this job.

Our eight-inch guns used to ice up so thick that the hydraulics wouldn't lift them. I used to go out there with another sailor, using a chisel and hammer to break the ice off the guns to lighten them. You also had steam hoses, if you could get them out. You had to wear a duffel coat on top of your overalls to protect yourself against the cold when doing such work. They gave us Long Johns made out of thick wool, but I didn't wear them. I sent mine home to my wife and she unpicked them and made something else.

Three stokers in the ice on *London's* upper deck. *W. Willing Collection.*

The 200 or so members of the engineering department rarely saw the outside world but, as Ted Huke recalled, sometimes they didn't need to:

I remember being on watch, seeing the heel indicator keep on going and stay there. I also remember going on the upper deck when the weather was really bad and it was thrilling, but I was too young to be worried. I was generally sea sick for the first three days at sea and thereafter I was okay, provided we kept going.

The world the *London's* crew inhabited could still yield moments of surprising beauty, as Norman Brigden recalled:

There were mountainous seas in the Denmark Strait – when we made six miles in twenty four hours – but at other times, when we were within 600 miles of the North Pole, the sea was very flat and the ship made a whispering noise when sailing through the sea slush. Then there was the biting cold of the Arctic when I was trying to de-ice some upper deck systems and also the glorious display of the aurora borealis.

Surgeon Lieutenant Ransome Wallis was not so convinced about the beauty of the Northern Lights:

The aurora borealis was a fearsome sight seen under these circumstances. I suppose in peace time one might have paid a lot of money to view this sort of thing but in war one was too conscious of the black and sinister Arctic Ocean which might easily be hiding a lurking U-boat. The unearthly brilliant rays of the aurora borealis only seemed to stress the finality of it all if disaster had struck.

The Surgeon Lieutenant recorded that, despite the dreadful conditions, the sailors somehow managed to keep in good physical condition:

In spite of a few cases of frostbite, the health of the ships' companies in the Arctic was extremely good. It was too cold for the ordinary pathogenic bacteria to flourish.

Keeping the minds of *London*'s crew healthy was a little more challenging, but the majority of the sailors had the strength of mind to ride out the storms.

A few people complained of psychiatric symptoms but there was no psychiatrist easily available and in most cases a few sharp words from the sick berth CPO accompanied by a No 9 pill containing a powerful purgative worked wonders. As I found out later in a destroyer, it was no good recognising the existence of such conditions as battle fatigue, fear neurosis or being bomb happy. There was nowhere to go to get away from it all so you just had to get on with the job. Probably most people had some degree of these conditions.

Able Seaman Tamon kept a secret diary of his time aboard the *London* that well illustrates the gruelling routine the cruiser endured in early 1942.

Feb 28. Left Hvalfjord for Denmark Straits – Patrol White.

This month 16 sea days, 1,931 miles.

March 7. Return to refuel, and leave to provide oil for destroyers convoying to Russia.

Met HMS Kent. Much ice on forecastle.

March 11. Arrived Iceland.

March 18. Left on Patrol White.

March 25. Returned Iceland. More ice.

March 30. Left on Patrol White. This month 20 sea days, 6,838 miles.

April 6. Returned to Hvalfjord.

April 11. Left on Patrol White. Very heavy seas.

April 18. Returned from patrol

Being a very junior sailor's point of view, it contained very little of the 'big picture' detail, reflecting instead the mind-numbing, bone-chilling monotony of enduring some of the worst seas known to man.

Notes

1. R. Ransome Wallis, *Two Red Stripes*.
2. Graham Bramley, in an interview with the author.
3. Benny Goodman, in an interview with the author.
4. D.K. Brown, *Nelson to Vanguard*.
5. ibid.
6. R. Ransome Wallis, op. cit.
7. Nicholas Monsarrat, *The Cruel Sea*.
8. ibid.
9. Ken Tamon, in an interview with the author.

Chapter Eleven

THE *TIRPITZ* CHASE

Poor Relations

Nothing illustrated the harsh reality of nearly three years of war better than the absence of what the new American allies regarded as commonplace but the Royal Navy's sailors saw as luxuries. In the desperate battle that was being waged in the Atlantic to get supply convoys through to Britain there was no room aboard merchant ships for anything that was not essential, and with the bill for fighting the war escalating every day, the nation could not afford them anyway.

Chatting with an American naval doctor, aboard a destroyer giving him a ride back to Hvalfjord after a visit to a US military hospital in Reykjavik, Surgeon Lieutenant Ransome Wallis listened with some amusement to the other man's concerns over his sailors' diet:

He told me that he was concerned that his men had not had any fresh grapefruit for two weeks owing to the non-arrival of the storeship and asked me if I thought they would get vitamin deficiency. I told him that I had not had any grapefruit for two years but felt all right.

The American naval doctor was shocked to hear this and promised to get some fresh fruit sent over to *London* as soon as the US Navy supply vessel arrived. The arrival of enough fresh vegetables and fruit to stock a greengrocer's provided a superb boost for the spirits of *London*'s sailors, who were wilting after their latest stint of Patrol White. Ken Tamon recalled:

It was fantastic. They sent over oranges, apples, grapefruit, lettuce and celery – such luxury. A couple of days later we left for another Patrol White, but at least we had some decent grub inside us.

Petty Officer Bramley was amazed at the generosity of the Americans, and the rich variety of their diet, when he rode down to Reykjavik on a US destroyer for some shore leave.

As we are going down the fjord, this bloke comes out of the galley carrying a tray with freshly cooked pork chops on it. By the time he got to the guardrail the chops had disappeared. He took it in good humour and said to us: 'I guess you guys are hungry!' So he let us into the galley and we cleaned up the fruit and any other food going. We never saw anything like it on a Royal Navy ship. The Americans lived like lords. We invited them over to our ship and we gave them bread and jam. When they got back to their ship they sent some more fruit over. They must have taken pity on us. The Americans had a never-ending supply – the American merchant ships at Iceland were stuffed full of goodies.

America's entry into the war was appreciated for much more than a welcome boost to the Vitamin C levels of *London*'s sailors. The US Navy made up for a serious deficiency in fighting strength of the Home Fleet at a time when the German threat in the Arctic was growing at an alarming rate.

Joining the battleship *Tirpitz* by the end of February in Norwegian waters was the pocket battleship *Admiral Scheer*, together with the *Prinz Eugen*. The latter was hit by a torpedo from the British submarine *Trident* as she approached Trondheim, and would return to Germany that May for repairs, but the pocket battleship *Lützow* and heavy cruiser *Admiral Hipper* also

HMS *Victorious* in an Icelandic fjord. *Strathdee Collection.*

migrated north. German surface warship strength was formidable; U-boats were operating from submarine pens in Norway and Luftwaffe airfields were being reinforced.

Against this build-up the depleted Home Fleet – sapped of strength by the need to replace ships lost in the Mediterranean and maintaining a presence in the Indian Ocean to counter the Japanese – was only able to field the battleship *King George V*, battlecruiser *Renown*, aircraft carrier HMS *Victorious,* several cruisers (including *London*) and a mixed bag of destroyers.

The new battleship, *Duke of York,* had only been completed in November 1941 and was on sea trials, while the other new King George V Class capital ships, *Anson* and *Howe,* would not be finished until the summer. In the meantime, the US Navy battleship *Washington,* heavy cruisers *Tuscaloosa* and *Wichita* plus a number of destroyers helped fill the gaps, joining the Home Fleet in April 1942. Prior to that the Americans were, of course, in the thick of it on the convoy runs from the USA to Iceland.

The US Navy's opinion of the Royal Navy was that it was verging on the antiquated. Aside from the living conditions of the ordinary sailors being inferior, the Americans saw that the Royal Navy's warships were creaking at the joints (literally in *London*'s case thanks to the mishandled rebuild) and their weapons and ammunition appeared inferior to US warships. Britain's Fleet Air Arm flew obsolete aircraft and the warship construction rate of the UK's yards was very poor compared with the USA's. In some quarters of the US Navy there was outright contempt for the British fleet, from admirals who carried on a tradition of enmity with its roots in the fight for freedom of the late eighteenth century.

Yet, there was something heroic about the Royal Navy that the American fleet, for all its wealth, could only dream of. The great battles of the Pacific that awarded the modern American fleet its spurs had yet to be fought and, in the wake of Pearl Harbor, while America was struggling to field a fully-fledged striking fleet anywhere, the Royal Navy was taking terrible losses in ships and men yet still managing to fight the good fight. Its sailors had the sort of grim determination the naive, unblooded Americans would only discover as their war against the Japanese gained in ferocity.

In the meantime, the US Navy was eager to help the British escort the supplies American industry was churning out for Russia to keep the pressure up on the Germans. The Royal Navy was no less eager than the Americans, but it appeared the easy days of convoys getting through to Murmansk and Archangel with little or no losses were drawing to a close.

In mid-February, HMS *London* became flagship of the First Cruiser Squadron (CS1) again, carrying Rear Admiral Louis Hamilton. Leading Telegraphist D.A. Hibbit was one of the Admiral's staff and he was not happy about joining the *London*, as he would have preferred to stay with *Norfolk*, the previous flagship, which was heading for a refit on the Tyne. The young sailor kept a journal in which he remarked, with some disgust, of his arrival in *London*: 'When we got aboard we discovered that there were no empty lockers to spare, so we had to keep our kit in our bags.'[1]

A Fleeting Chance

The big worry, above all others, was the *Tirpitz* and in early March 1942, before the American heavy ships arrived, the Home Fleet made a determined attempt to eliminate her.

On 1 March, convoy PQ12, consisting of sixteen merchant ships, had set sail from Iceland, with QP8, made up of fifteen supply vessels, coming in the opposite direction from the Kola Inlet. They were given the initials PQ (for Russian bound) and QP (for UK bound) in honour of a naval officer who helped organize them in the Admiralty, Commander P.Q. Roberts. A Luftwaffe scout plane sighted PQ12 on 5 March, when it was less than 100 miles south of Jan Mayen Island and the Germans decided to mount a foray by the *Tirpitz*. Hitler, who was keeping his heavy ships on a tight leash following the loss of *Graf Spee* and *Bismarck* and the damage to the *Prinz Eugen*, reluctantly gave permission. So *Tirpitz*, with a trio of destroyers as her screen, left Trondheim and headed north.

HMS *London* had returned to Hvalfjord from a Patrol White on 7 March and was taking on fuel. Leading Telegraphist Hibbit's journal recorded a dramatic development: 'At about

A grim silhouette – *London*'s distinctive one-off lines. *Strathdee Collection.*

Tirpitz **in a Norwegian fjord.** *US Naval History and Heritage Command.*

23.00 the same day we put to sea under urgent orders of the Admiralty. Travelling all night at high speed.'

Shortly after 10 a.m. the following day Captain Servaes went on the public address system to tell the crew what was happening:

Sometime yesterday afternoon, one of our submarines sighted the German battleship Tirpitz, *sister ship of the* Bismarck, *under way off the coast of Norway. The* Tirpitz *was reported to be steering in a northerly direction. It is thought that she is endeavouring to intercept an important Russian convoy of ours. The* Duke of York, *and* Renown *are already on their way to join the convoy and we, together with the* Kent, *and the* Sheffield *have been ordered to proceed to a position near the convoy, where we shall rendezvous with the escorting destroyers for the twofold purpose of oiling the destroyers and going into action against the* Tirpitz, *should our assistance be required.*[2]

The British submarine that spotted *Tirpitz* was the *Seawolf*, loitering off Trondheim and her sighting report allowed Admiral John Tovey, the Home Fleet Commander-in-Chief, to steam hard for an interception somewhere south of PQ12's position.

Most important of all were the Albacore torpedo-bombers of *Victorious*. But progress was slow, as the weather was dreadful. It was so bad that British scout planes could not find the *Tirpitz* to fix her for a strike and the Germans failed to locate PQ12. Giving up, Tovey sailed his force south to a position parallel with Trondheim, hoping *Victorious* would be able to mount a strike as the *Tirpitz* returned to her lair. At dawn on 9 March, the carrier launched two of her aircraft on an air search, with the scouting aircraft discovering the German battleship sixty miles off the Norwegian coast, heading for the Vestfjord. At around 9 a.m. a British strike force of more than a dozen Albacores arrived and scored two hits on *Tirpitz*, for the loss of two aircraft.

Unfortunately the torpedoes were defective and, with the Home Fleet's battleships still too far away to intervene, the *Tirpitz* was able to seek refuge in the Vestfjord unharmed. While the Home Fleet failed in its attack on the *Tirpitz*, at least both PQ12 and QP8 got through with light casualties and Hitler was so alarmed by such a close call that he barred any further sorties by *Tirpitz*, unless the British carrier had been located and destroyed.

Rear Admiral Hamilton corresponded regularly with his mother while he was aboard *London* and, in a letter dated 14 March, gave his own thoughts on the Home Fleet's vain attempt to sink the German battleship.

We have been doing a lot of sea-time and joined in the Tirpitz *hunt the other day. That ship is an infernal nuisance and the most important business of the war at present time is to cripple or destroy her — it would just make the whole difference in the way of freeing ships for other theatres.*[3]

Leading Telegraphist Hibbit told his journal the *Tirpitz* chase had been 'very unsuccessful' and he was still deeply depressed about being aboard the leaky old *London* rather than his beloved HMS *Norfolk*. He noted on 11 March:

...arrived back at Iceland...Very nice weather all following week — Just like summer but unfortunately cannot take any photos as no films can be obtained aboard this rotten ship.

Unfortunately for Leading Telegraphist Hibbit, Rear Admiral Hamilton rather liked having HMS *London* as his flagship, deciding to stay aboard even after *Norfolk* returned from refit. In a letter dated 26 March, Rear Admiral Hamilton wrote:

I have decided to stay in this ship as she is very good, and also 'Norfolk' *has the Senior Captain in the squadron. I am old fashioned enough to like a junior Flag Captain.*

The Strain Begins to Show

Like many officers in the Home Fleet, Rear Admiral Hamilton was frustrated by the lack of assistance from the RAF. Its Bomber Command was mounting 1,000-aircraft raids on German industrial areas and centres of population. However, Coastal Command was being starved of the aircraft it needed to mount long-range anti-submarine patrols and German warships were sneaking from the Baltic to Norway without being spotted by the RAF. The

Germans had been allowed to build submarine pens in Norway with no significant RAF response and, aside from three totally ineffective high-level raids early in 1942, no serious effort was being mounted to destroy *Tirpitz* in her fjord. The Royal Navy would have carried out the job with relish itself, but the lack of resources devoted to the Fleet Air Arm between the wars had left the British navy with not enough carriers to form a strike group in northern waters, nor with a dive-bomber suitable for the task. Admiral Tovey ruffled plenty of feathers in the Air Ministry and the Government by writing a number of scathing letters about the RAF's selfish and pig-headed conduct. The truth was that many naval officers also found the RAF's strategy simply barbaric.

Rear Admiral Hamilton revealed to his mother in a letter, written shortly after joining HMS *London*, that he felt it hopeless to believe '...we could win the war by bombing German women and children instead of defeating their Navy and Army.' In his letter of 26 March he returned to this theme:

> *I am afraid I feel that there is a grave risk of losing the war unless it is realised:*
> *1) That it is primarily a maritime war.*
> *2) That wars are won by killing the enemy's soldiers or sailors.*
> *3) The aeroplane is just as much a naval weapon as a submarine or destroyer.*

In the meantime, as Bomber Command sucked in vast resources to build the fleets of planes needed for its campaign against German cities and industry, the Royal Navy gritted its teeth and took all the punishment the Arctic could throw up.

HMS *London* was again beginning to buckle under the strain and one man who knew her leaks and other structural problems very well during this period was Lieutenant Commander John Pearsall, the cruiser's Senior Engineer, who had joined the ship in October 1941 during her refit.

> *She was a comfortable ship inside, but not particularly reliable. As a result of all the modifications she leaked like a sieve...water was always coming through into the oil tanks, contaminating the oil tanks. That was the main problem. We had at sea with us for some time a Constructor who was specifically there to suggest the corrections and necessary repairs...*[4]

Keeping the ship going in the Arctic was not the only challenge, for, as Lieutenant Commander Pearsall admitted, the sailors themselves could be a bit tough to handle:

> *We had some problems with the ship's company but they were on the whole very good. Of course during the war one had a number of hostilities only ratings who had probably rather shaky pasts, but generally speaking once you gave them some responsibility they were exceptionally good. It was quite a traumatic experience to come from civilian life and into a ship under war conditions with very little leave for them and a lot of discomfort... conditions on the mess decks were very often unpleasant. And also of course many of them had families at home in places that were being bombed... and they had losses in their families when they were at sea. They got the news when they got back into harbour and it was not very nice. Occasionally when they got leave there was somebody who went AWOL, but nothing more than usual. We did quite a lot of Patrol Whites... and I think one of the main problems in that ship was the effect of the weather, particularly in the winter. It was thoroughly unpleasant... very rough and could be very cold, and a large ship like* London *rolled very badly much of the time. Apart from water getting into the oil, occasionally in some compartments you would get bits of spare gear – machinery – breaking adrift... that type of thing. On one occasion, in a compartment*

that contained lubricating oil tanks, some spare gear broke adrift and pierced the tanks so the compartment became full of oil several feet deep.

The German Navy had its own problems with oil. On top of Hitler's restrictions on battleship and heavy cruiser movements, there was a shortage of fuel. The Germans therefore restricted themselves to launching eight U-boats, three destroyers and Ju88 bombers against PQ13 in late March.

A storm hit the convoy, scattered the ships and later, while the light cruisers HMS *Nigeria* and HMS *Trinidad* were desperately trying to gather them back together, the Germans sank two merchant vessels. A fight developed between *Trinidad*, supported by the destroyer *Fury*, and German destroyers. Another Royal Navy destroyer, the *Eclipse*, together with Russian warships, soon became involved. *Trinidad* fired a torpedo that went haywire, because the cold interfered with its gyros. The cruiser's sailors were horrified to see the weapon coming straight back at them. Sustaining serious damage, *Trinidad* limped towards the Kola Inlet. U-boats now pounced on the convoy, claiming three merchant ships, but *Fury* sank one of them – *U-585* – while she was trying to finish off *Trinidad*. A German destroyer – *Z26* – was sunk in the fighting while several British escort ships were badly mauled.

Bad weather kept the German surface ships in harbour and hampered the U-boats and enemy planes but they still took their toll. Of PQ14's twenty-four ships, only seven got through to Russia – one fell victim to a U-boat while sixteen turned back due to bad weather. Of QP10's sixteen Iceland-bound ships four were lost to German attacks, which saw the advent in greater numbers of specialist torpedo-bombers.

On 28 April HMS *London* left Scapa in company with *King George V*, the USS *Washington*, USS *Wichita*, USS *Tuscaloosa* and HMS *Nigeria* together with HMS *Victorious* and a brace of destroyers. This Home Fleet task force patrolled between Norway and Iceland to cover PQ15 and QP11. The latter left the Kola Inlet on 28 April, with a close escort led by the cruiser HMS *Edinburgh*.

The American battleship USS *Washington*, reinforced the Royal Navy's Home Fleet in the spring of 1942. *US Naval History and Heritage Command.*

On 29 April the pocket battleship *Admiral Scheer* was reported to be at sea but she turned around without attacking the convoy or giving the Home Fleet enough time to find and sink her. The heavy escort force linked up with PQ15 on 30 April and the following day the Germans made a determined attack on the cruiser squadron. Because the cruisers' job was to hover just over the horizon, waiting for a surface warship threat rather than providing close anti-aircraft cover, the Luftwaffe's bombers ignored them, choosing to concentrate on the merchant ships. But on 1 May, the cruisers, and *London* in particular, came in for some special attention. Action stations were called at 10 a.m. and the *London*'s sailors dutifully went to their allocated positions throughout the ship.

> *...and the Captain's voice came over the loudspeaker almost at once saying that there was a wave of German dive bombers approaching and that we were about to be attacked... He had hardly finished speaking when the twin 4-inch guns on the port side over our heads opened fire and we could hear multiple pom-poms banging away. A voice over the loud speaker then shouted, 'we've got one' and then there was silence. A few moments later I heard the sound with which I was later to become so familiar, the clanging on the ship's side as if a giant were hitting it with a sledge-hammer. On this occasion it was quite a gentle thud for the bombs exploding in the sea were quite a distance away. Then the starboard 4-inch opened fire for a short time, then silence again except for the usual creaking noises of the ship. My party sat silently and rather tense and stared at the deck... .[5]*

After the order to secure from action stations was broadcast, Surgeon Lieutenant Ransome Wallis was among those who went up onto the upper deck to see what was happening. Although it was shortly before midnight, there was still bright 'daylight'. He could see *Nigeria* nearby and a destroyer following on behind. He asked some of the AA gun crews what had happened. They told him that six Ju88 dive-bombers had come in to attack *London*.

> *...and the leader peeled off into his long shallow dive apparently aiming at* London *but we shot him down before her could do any damage and the aircraft crashed into the sea just ahead of HMS* Nigeria *who ground it underfoot. A man was seen in the sea but this was no time to stop to pick up survivors. The cold would soon finish him off. Anyway he had tried to kill us a few minutes earlier.*

The other German bombers had veered away to unload their bombs on merchant ships. The same day, the *King George V* rammed one of her escorting destroyers, and was damaged by the smaller warship's depth charges going off as she was crushed.

Meanwhile, *Edinburgh* had been torpedoed on 30 April but survived and turned back for Murmansk, but on 2 May her luck ran out when three German destroyers attacked, one of them putting another torpedo into her. British destroyers came to the cruiser's aid but she was so badly damaged by now that she had to be abandoned and scuttled. The German destroyer *Hermann Schoemann* was also scuttled, having suffered fatal damage during the fight with *Edinburgh*. In the meantime QP11 linked up with *London* and the other Home Fleet ships, managing to get through to Iceland with the loss of only one of thirteen merchant ships. Three of PQ15's merchant ships were lost, with the other twenty-two reaching Murmansk.

HMS *London* in action on an Arctic convoy amid high seas. *Dennis Andrews.*

HMS *London* was back at Scapa Flow by 5 May and nothing much happened for nine days until, as Able Seaman Ken Tamon's secret diary recorded, there was a bit of a flap:

> *May 14-15. 3 am panic stations. Left with* Duke of York, Victorious *and flotilla of destroyers, etc.*
>
> *May 16. Met USS* Washington *and* Tuscaloosa *to cover* Trinidad *limping home from Russia.*

However, their mission was already in vain. After temporary repairs in Murmansk, *Trinidad* set out for the UK on 13 May, with a close escort provided by four destroyers, and four other cruisers providing more distant support. But German aircraft caught *Trinidad* late in the evening of 14 May. Following at least two bomb hits which caused an uncontrollable fire, *Trinidad* was abandoned, with one of the escorting destroyers sending her to the bottom with a torpedo in the early hours of 15 May.

The loss of *Edinburgh* and *Trinidad*, together with the growing numbers of torpedo-bombers and undiminished German surface threat, persuaded the Admiralty that convoys should be suspended during the perpetual daylight of the Arctic summer.

But, Russia was still under the most serious pressure from the Wehrmacht, which was preparing to mount a big thrust into the Caucasus towards Stalingrad. Also, more and more American merchant ships packed with essential war supplies were queuing up to come across to Iceland on the way to Russia, with President Roosevelt telling Winston Churchill there could be no delays.

Notes

1. Papers of D.A. Hibbit, Imperial War Museum.
2. ibid.
3. Papers of Admiral Sir Louis Henry Keppel Hamilton, National Maritime Museum.
4. John Pearsall, Sound Archives, Imperial War Museum.
5. R. Ransome Wallis, *Two Red Stripes*.

Chapter Twelve

THE SHAME OF CONVOY PQ17

America Piles on the Pressure

Convoy PQ16 in late May was composed of thirty-five ships, with seven of them lost to German action, while one was forced to head back to Iceland. Compared with other convoys this was a high attrition rate. Although there was twenty-four hour daylight, the ice barrier was at least receding north, giving the convoys more room to evade the Germans. The *Tirpitz* and *Hipper* were at Trondheim along with a quartet of destroyers, while at Narvik there was the *Scheer* and *Lützow* and around half a dozen destroyers. They failed to venture out and left the Luftwaffe and submariners to it. Luckily QP12 got through unscathed, with one merchant ship returning to Russia and the other fourteen reaching Iceland.

At the time there was heavy pressure on the British from the Americans to mount a joint invasion of France. The British considered this to be impossible. The situation in North Africa was still volatile, with Rommel storming across Libya to threaten Cairo and the Suez Canal. Explaining that their overstretched forces could not handle yet another front, the British managed to persuade the Americans to help them finish off the Germans and Italians in North Africa as a precursor to invading Europe. Having agreed to this strategy the Americans were not about to concede a temporary cessation of the Arctic convoy runs to Russia, with over 100 of their merchant ships now ready to make the voyage.

While the Russians have never given a completely honest account of how much they relied upon the weapons and other supplies that came via Murmansk and Archangel, to have turned off the supply at this point would surely have been a serious blow. In late May 1942, the Germans launched their spring offensive, defeated a Russian counter-attack at Kharkov and got ready to send the Sixth Army deep into the Caucasus. Although the transfer of Soviet weapons factories to the Urals had been carried out successfully, with thousands of artillery guns and new T34 tanks rolling off the production lines, the situation was still precarious. The cargo carried by convoy PQ17, set to depart Iceland at the end of June, was precious indeed. Six hundred tanks were being carried to Russia, along with more than 3,000 assorted vehicles (including trucks and ambulances), 210 aircraft, plus over 100,000 tons of foodstuffs, ammunition, radar equipment and medicines.

American supply ships that had reached Iceland were the lucky ones, for the US Navy's decision to send most of its destroyers to the Pacific, and its abject failure to introduce a convoy system, had allowed German U-boats to sink 492 Allied ships off the east coast of the USA between January and late June. Senior officers in the US Army lambasted the top brass of the US Navy for failing in their duty, claiming that the whole war effort was under threat.

Having survived the perils of home waters, the American merchantmen arriving in Iceland for PQ17 had also run the trans-Atlantic gauntlet. In June 1942, the score sheet for the Allies in the Battle of the Atlantic was depressing in the extreme. U-boats sank 128 ships totalling 650,000 tons, the majority of them lost in waters off the USA. The Germans lost just two submarines.

USS _Tuscaloosa_, in the foreground, and HMS _London_ at Scapa Flow, summer 1942.
US Naval History and Heritage Command.

The anger and despair was so great aboard American merchant ships that armed US Navy officers were attached to their crews to ensure they didn't mutiny or abandon ship. Of the thirty-six merchant vessels assembled for PQ17, twenty were American, eight were British and two were Russian together with one each from Norway and Holland. Three Merchant Navy rescue ships were part of the convoy, along with a fleet oiler for the warships.

The close escort, under Commander Jack Broome, consisted of six destroyers, four corvettes, two anti-aircraft ships, three minesweepers and four armed trawlers. Broome commanded his group from the destroyer HMS _Keppel_.

American and British warships assembled in the Hvalfjord in the summer of 1942, including HMS _London_, whose distinctive lines can be seen in the middle distance on the right of the picture. HMS _Norfolk_ is the warship in the foreground _US Naval History and Heritage Command._

The cruiser escort was provided by the First Cruiser Squadron, with *London* as flagship alongside *Norfolk*, and the US Navy's *Tuscaloosa* and *Wichita*. They were accompanied by three destroyers, HMS *Somali*, USS *Wainwright* and USS *Rowan*. Providing distant cover, in case the German heavy ships ventured forth, was the Home Fleet, led by the *Duke of York* with *Victorious* and the American battleship *Washington*. It was the first time the Americans had taken part in actively covering a Russian convoy and both warships and merchant vessels involved in PQ17 were under British command.

The convoy left Iceland on 27 June, the day before the Germans began their thrust into the Caucasus. Two of the American merchant ships dropped out early, after one ran aground and the other hit ice. The Luftwaffe's scout planes kept up a permanent watch on PQ17 and German naval intelligence tracked the convoy's position using radio intercepts.

By 1 July the convoy was past Jan Mayen Island, with the first U-boats sighted on the same day and attacked by escorting destroyers. The cruiser force was keeping just over the horizon where it was enthusiastically engaged in depth-charging submarine contacts. The cruisers' aircraft, including *London*'s Walrus seaplanes, were sent up on Anti-Submarine Warfare patrols to provide more air cover. The weather was calm, with fine blue skies, but there were the usual patches of fog, caused by the warm Gulf Stream hitting the cold polar water and random ice.

The Wolves Gather

German spies had been able to pass on considerable detail about PQ17's composition and a pack of eleven U-boats had been assembled to attack it. In fact, the Germans were planning something special – a concerted offensive involving U-boats and bombers backed up by a heavy ship sortie. Codenamed *Rosselsprung*, or Knight's Move, it had been outlined for Hitler on 1 June, when he was told it would involve *Tirpitz, Lützow, Hipper* and *Scheer* attacking PQ17 once it reached the Barents Sea around 5 July.

The German Naval Staff was confident of evading the Home Fleet and anticipated great success, but Hitler reminded his admirals that HMS *Victorious* had to be found and destroyed before the battle group could leave harbour. But, if the big ships remained where they were the plan would be totally unfeasible – they would never catch up with the convoy. So the Kriegsmarine persuaded the Führer to allow *Tirpitz* and *Hipper* to move up to the Vestfjord, and *Scheer* and *Lützow* to the Altenfjord, in the far north of Norway, prior to the green light being given. The admirals also persuaded Hitler the operation could proceed in safety if the *Victorious* and Allied battleships had been confirmed as too far away to intervene before the

A German aerial reconnaissance image of the convoy PQ17 while still under escort on the way to the Soviet Union. *US Naval History and Heritage Command/ NH 71382.*

U-251 returns to Narvik after attacking convoy PQ17. *US Naval History and Heritage Command/ NH 71312.*

convoy was destroyed. However, during the move north on 3 July the *Lützow* ran aground, as did three destroyers escorting *Tirpitz*. Robbed of a pocket battleship and left with a reduced escort force, the Germans decided to concentrate their battle group in the Altenfjord.

Meanwhile PQ17 was passing just to the north of Bear Island, U-boat contacts were coming in thick and fast and air attacks were growing in intensity. At 3 p.m. a wave of two dozen German bombers came over but, because of a layer of low cloud, the attackers were forced to drop their deadly payloads through it and all the bombs missed. A U-boat fired three torpedoes at the *Tuscaloosa*, but they all went wide. In the early hours of 4 July 'a single plane shot out of a fog bank...and dropped its fish'[1] hitting a cargo vessel and damaging her so badly she had to be scuttled.

In the midst of the fighting there was still time for important national celebrations. To celebrate Independence Day, *Tuscaloosa* and *Wichita* put up ceremonial bunting. Rear Admiral Hamilton sent the American ships a message of congratulation:

> *On the occasion of your great anniversary it seems most uncivil to make you keep station at all, but even the Freedom of the Seas can be read in two ways. It is a privilege for us all to have you with us and I wish you all the best of hunting.* [2]

The *Wichita* replied that she was happy to be part of Hamilton's command.

At 5.15 p.m. a gaggle of torpedo-bombers came winging over smooth seas and sank another merchant ship. Just under an hour later more than thirty German aircraft were greeted with a storm of fire from the escorting warships and merchantmen. Several were shot down but two more ships were mortally wounded.

London contributed to the air defence with her 8-inch guns, helping to disrupt an attack by Heinkel torpedo bombers, but on the whole the cruisers had to content themselves with being spectators.

Over in the cruiser force they watched the battle from less than 10,000 yards away, still keeping to instructions and staying out of range of concentrated attack and the constant U-boat danger – waiting for the surface situation to be clarified by the Admiralty. All fretted in their inability to help as they gazed on a sky packed with brown and black blotches of bursting shells. London and Norfolk had set up a long-range barrage, engaging the enemy formations briefly as they flew past, though the aircraft paid little heed to the cruiser squadron and headed straight for their primary target, the convoy. The fact that London *managed to get one plane just as it was going over the horizon was poor consolation.*[3]

The US cruisers' shells were not of the air burst variety so they could not contribute – their shells had to achieve a direct hit to cause any damage.

Petty Officer Bramley had played a key role in enabling *London* to shoot down enemy aircraft with her main guns:

I devised a new way to set the fuse. I made this adaptation to the fuse setting cap so it could be set by hand for a certain height. As you can imagine the 8-inch gun had a much larger shell than our AA 4-inchers. The 8-inch shell created a bigger explosion and the gun that fired it had a longer range, so when it got near a German plane it did a good job.

By the evening of 4 July the Admiralty was confident that *Tirpitz* was in the Altenfjord and it also knew about the imminent foray, thanks to intelligence sources in Sweden. The increasing volume of German signals traffic seemed to indicate that something was about to happen and in the early hours of 4 July the Admiralty had advised Rear Admiral Hamilton, aboard *London*, that he should stay with the convoy until it was level with North Cape (25° East).

During the convoy's planning stage Admiral Tovey had suggested that once the German battle group was at sea PQ17 should be turned around to lure it into range of the Home Fleet. The First Sea Lord, Admiral Sir Dudley Pound, would not allow this and his orders stated that the cruisers should turn around at Bear Island, unless the convoy was attacked by German destroyers that could be effectively countered by the intervention of Hamilton's ships. Aside from avoiding major surface units, the Anglo-American cruiser force was also to ensure it did not stray too deeply into areas where U-boats were most active. Tovey

German battleship *Tirpitz* in the Altenfjord. *US Naval History and Heritage Command/NH 71390.*

instructed Hamilton that any attack by the German heavy ships was only to be shadowed, so that the Home Fleet could be brought to bear. The cruisers were only to attack if fog gave them extra protection, or if an air strike by Albacores from *Victorious* had inflicted such damage that the odds had been levelled. Additionally the standing Atlantic Convoy Instruction implemented in April 1942 'specifically forbade scattering of a convoy "until or unless the escort is overwhelmed". And on 21 June Tovey had ordered that Arctic convoys were to follow ACIs.'[4]

The Admiralty's final orders for PQ17 had ultimately incorporated both the scatter option and the possibility of it being turned back to lead the German surface ships within striking distance of the Home Fleet. On the afternoon of 4 July, Admiral Tovey, now with the Home Fleet somewhere between Iceland and Bear Island, advised the First Cruiser Squadron that it could leave PQ17 before reaching 25° East, indicating that the threat from major German surface units was receding. Hamilton replied that he would leave the convoy late on the evening of 4 July after refuelling his destroyers. Shortly before 7 p.m., *London* received a signal from the Admiralty instructing the First Cruiser Squadron to stay with the convoy.

Rear Admiral Hamilton was not impressed by the contradictory attitudes of his two masters, but he knew that, even though Tovey was the man on the spot with excellent operational instinct, the Admiralty was tapping in to a steady flow of intelligence. He felt that the best course of action would be for his squadron to stay with the convoy until the exact disposition of the German surface ships was known. Despite instructions not to attack major German surface ships, Hamilton had decided that he would have a go if a heavy cruiser or pocket battleship came over the horizon. If his squadron encountered either *Scheer*, *Hipper* or even *Lützow* on their own, he would split his force and attack from four directions – a tactic similar to that employed by the cruisers *Exeter*, *Achilles* and *Ajax* against the *Graf Spee* in 1939. If there were two German heavy ships then *London* and *Norfolk* would attack one, while the Americans would attack the other. If *Tirpitz* appeared then *London* and the other cruisers would, as instructed, hang back and try and lead her towards the Home Fleet. As Hamilton pondered these options in his cabin aboard *London*, the convoy was coming under sustained attack from German aircraft.

The cruisers were five miles in front, zig-zagging to reduce the risk of being hit by torpedoes, but as the enemy aircraft were concentrating as usual on the merchant ships, the *London* relaxed her state of alert.

Watertight doors remained clipped, some damage control teams stayed at their action stations and, of course, the gun crews stayed closed up, but anyone not required was allowed to go up on the upper deck and see what was going on. Engineering Artificer Bruty was one of those who managed to get some fresh air:

> Because we were at Action Stations there was an extra artificer in each boiler room and I went up on deck for a bit of a skive, having told the Chief I was going to B boiler room, or something like that. It was then that I saw this German submarine on the surface. We didn't stop, we just carried on steaming and fired a salvo at it as we passed.

That evening a British and an American merchant ship were sunk and a Russian tanker damaged, with three German aircraft shot down.

At the Admiralty in London the latest deciphered German signals revealed the enemy had mistaken HMS *London* for a King George V Class battleship, sailing in company with three escorting destroyers. Because of her one-off rebuild, the *London*'s silhouette was like no other cruiser's and, unless scout plane crews were warned of her presence with the convoy, it was an easy mistake to make. The First Sea Lord was briefed by a member of his staff about the likely implications of this mistaken sighting and other factors that might create caution in the Kriegsmarine's high command. It was pointed out that the PQ12 experience would have made the Germans very wary of risking *Tirpitz* at sea without knowing for certain where *Victorious* was. In fact, unknown to the British, floatplanes from the American cruisers had been mistaken for Albacores from the British carrier. However, the person who briefed the First Sea Lord did know that Royal Navy and Russian submarines staking out the Altenfjord had not seen the *Tirpitz* emerge. But, when Admiral Pound asked for a cast iron assurance that the *Tirpitz* was still in the Altenfjord it could not be provided.

> *Pound, forced to act on the basis of a worst-case scenario – that the* Tirpitz *was at sea and already closing on PQ17 – came to the fateful decision. The convoy was to disperse.*[5]

Admiral Pound began to write the first of three signals that would hit the convoy like a bolt out of the blue.

> *...only four ships had been sunk after heavy air attack and the convoy was trundling happily along in the calm weather between Bear Island and pack ice with two thirds of its distance run... As day melted into endless twilight there was a lull. Even the enemy's shadowing aircraft withdrew for a spell. Suddenly the Arctic's peace was shattered. Not by the enemy, but by three signals which arrived from Admiralty in quick succession.*[6]

On the bridge of HMS *London*, Rear Admiral Hamilton was suggesting that the break in enemy attacks provided an opportunity for a quick bite to eat. Feeling in a good mood, having brought the convoy through what was probably the worst of the action they would have to endure, *London*'s officers suggested some humour might also be in order.

> *...the Admiral approved the suggestion of hoisting Pendant 7 which still has an additional meaning that 'Flag and Commanding Officers have time for the next meal...'.*[7]

Within seconds of the food arriving on the cruiser's bridge the first signal arrived. Marked 'MOST IMMEDIATE' it said:

> *CRUISER FORCE WITHDRAW TO THE WESTWARDS AT HIGH SPEED.*[8]

The time was 9.11 p.m. and twelve minutes later another equally alarming signal hit the *London*.

> *IMMEDIATE... OWING TO THREAT FROM SURFACE SHIPS, CONVOY IS TO DISPERSE AND PROCEED TO RUSSIAN PORTS.*[9]

This was followed, at 9.36 p.m., by the signal that has gone down as one of the most notorious in naval history:

> *MOST IMMEDIATE. CONVOY IS TO SCATTER.*[10]

Such a rapid succession of signals, rising in intensity, could mean only one thing – the *Tirpitz* and her consorts were about to appear over the southern horizon. All good humour drained away:

> *Away went the meal; down came Pendant 7 to be rammed ignominiously back into its locker with strict orders from the Admiral it was never to be used thus in his Squadron again.*[11]

Rear Admiral Hamilton exchanged signals with Commander Broome in the *Keppel*. They agreed that the six close escort destroyers should join up with the cruisers for the charge west against the German capital ships. In the USS *Wichita* Lieutenant Douglas Fairbanks, the movie star, thought the 'Scatter' order must be a sick joke:

View from a U-boat as the submarine comes alongside lifeboats from the American steamer *Carlton*, torpedoed on 5 July 1942 after PQ17 convoy was ordered to scatter.

US Naval History and Heritage Command/NH 71305.

Our reaction was one of stunned shock. We felt there must have been some error in transmitting the signal.[12]

And, contrary to Hamilton and Broome's assumptions, the Americans believed the *Tirpitz* was coming at them from the east and the British were running away. They were:

...particularly bitter, cursing the British for what they believed to be running away in the face of a good battle which we had a chance of surviving...Our anger was made more intense by the philosophic and good-natured spirit in which the merchant ships received the order and saw us turn tail.[13]

In one of the Royal Navy AA ships sailors watched the cruisers disappear over the horizon and felt a '...a moment of awful emptiness'.[14]

Aboard the American merchant vessel *Daniel Chase*, the skipper saw the destroyers joining the cruisers as they passed through the convoy and cursed them: '...the bloody Navy has left us bare-assed.'[15] But, far from obeying the standing orders not to pick a fight with the bigger German warship, Hamilton and Broome were now determined to do their best, in the fine traditions of the Royal Navy, to protect the convoy and slow down the *Tirpitz* enough for the Home Fleet to destroy her.

As the *London* picked up speed, Gordon Bruty guessed that another big chase was on:

...there was a lot of activity. We were put on Action Messing and then Action Stations. The speculation rose that we might be going into action against a German surface ship. You could feel the London *turn around and we were suddenly at full speed. Then the buzz came down that we are going up against the* Tirpitz. *We were expecting to go into action at any moment.*

Surgeon Lieutenant Ransome Wallis listened intently as the Admiral used *London*'s public address to tell the crew what he believed was happening:

The impression we got and which he certainly believed himself was that we were about to do another Jervis Bay *action or perhaps, even better, to put up a successful fight against the odds.*

Those sailors who had been sent to *London* after surviving the sinking of the *Jervis Bay* no doubt scoffed at the idea of PQ17 escaping destruction in the same manner. The circumstances were fundamentally different to those experienced in November 1940. For a start, when the convoy the Armed Merchant Cruiser *Jervis Bay* was escorting, H.X. 84, was scattered, the *Admiral Scheer* was in sight and soon opened up with her guns. The decision to scatter was taken by the skipper of the *Jervis Bay*, Captain Edward Fegen, who then attacked the *Scheer* fully aware that his seven 6-inch guns were no match for the German pocket battleship's six

11-inch main armament, or her eight 5.9-inch guns. This was all well and good, and *London* with the other warships could certainly put up an equally good fight before succumbing to *Tirpitz*, possibly even scoring heavy damage. But, in November 1940 H.X. 84's ships were able to disappear in four directions – there was no ice sheet to the north that forced them to head south, closer to the enemy. They were able to head for friendly ports on all sides, with Canada not far away. In the vastness of the Atlantic there were no massed bomber attacks possible to assist the *Scheer's* depredations. There were no U-boat wolf packs lurking nearby as part of a coordinated plan, for the *Scheer* was a lone wolf. Thirty-one out of thirty-seven ships made it to safety. Surgeon Lieutenant Ransome Wallis thought scattering a convoy in the Arctic made no sense at all:

> *It had been shown over the past nine months that in these waters the only real chance of safety lay in rigid station keeping by the merchant ships and above all 'no straggling'. Ships that straggled were picked off by U-boats or aircraft with great ease and regularity.*

However, in the Barents Sea, as the midnight sun bore down on the night of 4 July, senior officers in HMS *London* and the other warships trusted that the Admiralty 2,000 miles away actually knew better than they did. The only initiative taken had been the decision to join the close escort destroyers with the cruiser force as it sped away through the convoy. With *Tirpitz* about to appear over the horizon it would have been madness not to, after all it took much of the Royal Navy to destroy her sister, the *Bismarck*.

Surgeon Lieutenant Ransome Wallis thought that with fog patches and smokescreens the nine destroyers and four heavy cruisers could put up a decent fight:

> *At any rate we girded our loins and prepared to do our best as the Admiral swung his cruiser force round between the scattering convoy and the direction from which the Germans were expected to emerge.*

A Fool's Errand

Keeping up a bone-jarring twenty-five knots the '...destroyers and cruisers continued a nightmare dash at high speed through fog and a sea scattered with icebergs.'[16] The ships kept a fairly tight formation and there was some risk of collision at high speed.

> *...the look out astern on* Keppel, *suddenly hearing many more voices than usual, lowered the binoculars with which he had been trying to peer into the mist just in time to see a blacker wall than the fog closing in on the destroyer. Very quickly it proved to be the port side of a much larger vessel and only a matter of feet away. The great ship drew away with the same speed as she had approached, the air then filling with the discordant music of sirens. Only by a fraction had* London *escaped sending* Keppel *to the bottom.*[17]

But as each mile slid by under the keels of their warships, with no sign of the German battle fleet, the British and American sailors grew increasingly uneasy. Surgeon Lieutenant Ransome Wallis gazed at the faces of the people around him in HMS *London* as they became ashamed and depressed.

> *...gradual disillusionment set in – a feeling so strong that it swamped the natural feeling of relief at being let off the hook. We carried on to the west at a fairly high speed; in the eyes of the merchant ships we were running away and we did not like it.*

As the cruisers and destroyers sped on, the communicators in *London* read with mounting horror the signals from PQ17 merchant ships as they were ripped apart by aircraft and U-boats. Commander Broome offered to go back but Rear Admiral Hamilton reminded him that his ships were low on oil and had lost their fuel ship somewhere in the scattered convoy.

In his secret journal Ken Tamon noted both the mood of the men and the faintly ludicrous attempts to buck them up with air attack exercises:

Ship's company morale at rock bottom. Commander Edden exercises ship's company in defence and damage control, impersonating bombs with 'CRUMPA!' over the tannoy.

Gordon Bruty felt equally overwhelmed:

We realized that we were a damp squib... looking for a phantom enemy fleet. Down in the engine room we didn't know they had been told to scatter or that they were being slaughtered, being sunk left, right and centre... but we did know that we had left them to it.

Captain Servaes picked up on how bad the men were feeling when he took a meal break during the endless night of 4-5 July.

I always lived on my bridge during these operations, and that night as I was having my evening meal on the bridge my old chief steward came up to me, and as he cleared away my plate and cutlery he whispered to me, 'It's a pity, Sir, that we had to abandon that convoy...'. I knew then that the story was round the whole ship, and that something would have to be done to bolster the crew's morale.[18]

Admiral Hamilton sent out a signal to all ships:

I know you will all be feeling as distressed as we are to leave that fine collection of ships to find their own way to harbour... I am sure we shall all have a chance of settling this score with the enemy soon.[19]

Far to the east the thirty merchant ships were trying to make it to wherever they could find shelter – some sought out the coves of Novaya Zemlya and others tried to make straight for Archangel or Murmansk. But, with only a few trawlers, AA ships and corvettes to fend off the U-boats and bombers they didn't stand a chance. In just five days – 5 July to 10 July – twenty PQ17 merchant ships were lost. Ultimately only eleven merchant ships reached their intended destination – two British, seven American and the two Russians. A number of American vessels gave up the fight, their crews abandoning ship while they were still seaworthy, preferring lifeboats to sharing space with thousands of tons of fuel oil or ammunition. Others were deliberately run aground so their crews could flee, often straight into the arms of the Germans.

In *The Destruction of PQ17* David Irving paints an even more unflattering picture. He tells of crews mutinying, ships running up surrender signals and claims that nine of the cargo ships were abandoned while in a seaworthy condition, some skippers scuttling their ships on purpose to improve their chances of survival. Irving does at least mention the heroism of ships that persevered and did make it through. One of them was the *Daniel Chase*, which fought valiantly, but succumbed to damage from bomb near misses and was finished off by a U-boat after her crew abandoned ship. Aboard one of the minor escort ships that had stayed with the convoy was Edward Reynolds, who had been a boy sailor aboard *London* during the 1930s. He witnessed one of the rescue ships taking a serious hit:

*...they were coming in from three directions at once. And one stick of bombs actually took the *Zamalek, or the *Zaafaran, right out of the water so that you could see daylight between the keel and the water.[20]*

As for the *Tirpitz*, she actually set sail from the Altenfjord fifteen hours after the order to scatter had been sent by the Admiralty. It was a tentative foray by the battleship, accompanied by the *Hipper*, *Scheer* and half a dozen destroyers. It was prompted by reports of PQ17's cruiser force being withdrawn and confirmation that the Home Fleet, including *Victorious*, was somewhere to the north-west of Bear Island.

The German battle group was steaming into the Barents Sea by the afternoon of 5 July, but it was tracked by Russian submarines and also aircraft. In fact, one of the Russian boats had fired torpedoes at the *Tirpitz* off the North Cape, but they all missed. The German naval high command – knowing that Hitler had given permission for the sortie with extreme reluctance – was still paranoid about losing *Tirpitz*, or having her heavily damaged. It appeared the objective was already being achieved by other means anyway, so why continue to risk the fleet flagship? They withdrew the battle group.

On the Brink of Mutiny

Meanwhile the Anglo-American cruiser force joined up with the Home Fleet, as Leading Telegraphist Hibbit, on Admiral Hamilton's staff, noted in his journal:

> *Monday 6th at about 05.00 G.M.T. in approximate position 72 deg N, 2 deg E, we* (London, Norfolk, Tuscaloosa *and* Wichita *and 9 destroyers) rendezvoused with H.M.S.* Duke of York *(Battleship) and the U.S. Battleship* Washington *also H.M.S.* Victorious, HMS Cumberland *(8") and HMS* Nigeria *(6") and several escorting destroyers. So that the complete force consists of 2 battleships, 1 aircraft carrier, 6 heavy cruisers and about 19 or more destroyers. It is a wonderful sight. They stretch from horizon to horizon.*[21]

Now Surgeon Lieutenant Ransome Wallis saw depression turn to fury: 'The mood of the *London's* crew was almost mutinous...'.

In the *Wichita*, the US Navy's sailors were equally angry, but found some consolation in the fact that their feelings were mirrored by the Royal Navy cruisers.

> *Only when we were clear of the immediate battle and steaming back in line ahead for Scapa Flow were we able to compare notes with the British cruisers, and found they were as bitter and angry as we were.*[22]

A special edition of the *Wichita's* newspaper had been published on 5 July, which, after explaining how serious the threat from the German heavy ships had been, said:

> *None of us likes the feeling of seeming to run away but it must be remembered that we are in possession of only limited information. No one can accuse us of ever having a faint heart...*
> *...Nor can anyone say the British lack guts.*

The paper pointed out to the crew that Britain had been fighting the Germans for three years – one year on their own – with their navy spread across the world, while the US Navy was only seven months into the war.

> *...war takes a lot of patience and broad thinking as well as action and courage.*[23]

London and *Wichita* followed *Duke of York* to Scapa, getting in on 8 July, while *Norfolk* and *Tuscaloosa* went back to Iceland. Convoy QP13 from Russia arrived there just before them, but unfortunately, having evaded the worst of German attacks, five out of the thirty-five merchant ships had been blown up after accidentally straying into a British minefield.

Recriminations

In a brief exchange between Hamilton and Tovey on the journey south, the latter had expressed surprise at PQ17's close escort destroyers being withdrawn too and indicated that the Admiralty would not be pleased. Hamilton realized that he could expect further criticism of this move and it would have far-reaching consequences. He wrote a pre-emptive letter to Tovey:

As things have turned out, I may be severely criticized by the Admiralty, but judging the situation on the spot, with action against superior force imminent, I feel my decision was correct.[24]

Hamilton also told Tovey that *London*'s lower deck was very upset about 'running away' from the convoy and, as soon as the cruiser arrived at Scapa, he went to see him. When Hamilton had been given the full facts by Tovey he was furious the Admiralty had in fact known no more than the man on the spot – they were guessing as much as he was. He felt betrayed.

That afternoon, he and his Flag Captain, R.M. Servaes, went ashore alone together to climb the hills bordering on the Flow. They exchanged little conversation until they had reached the top, and stood gazing out over the Home Fleet at anchor. Then Hamilton said, 'Well I suppose I ought to have been Nelson. I ought to have disregarded the Admiralty's signals.' Captain Servaes shook his head and said that even Nelson could not have ignored a series of signals like those.[25]

It was true. Signal flags flying from Hyde Parker's HMS *London* at Copenhagen could easily be ignored by the half-blind Nelson in the heat of battle, with no fear of any further interference. But in twentieth century warfare, with direct links to the Admiralty and no enemy ships in sight, Hamilton had no leeway to ignore stupid orders, particularly when they came in threes.

Back aboard *London*, Gordon Bruty felt equally depressed:

By the time we got to Scapa everyone was feeling dejected. We knew we had gone on a fool's errand. By this time we realised many of them had been sunk trying to get to Archangel on their own.

Hamilton decided to set the record straight with *London*'s sailors by addressing them on the warship's quarterdeck. First he warned them not to misconstrue his comments as criticism of the Government or the Admiralty. He also advised them not to mention the PQ17 disaster in letters home. The Admiral then explained as much as he knew about the circumstances surrounding the scatter order, reiterating the threat from the German heavy ships. But, he could not fail to let them know how bad he felt at abandoning the convoy:

I have never hated carrying out an order more in my life.

The sailors cheered. Hamilton went on:

I felt – as I know all of you felt – that we were running away and leaving the convoy to its fate. If the decision had been left to me, I would have stayed and fought – and I should have been wrong. You have got to put personal feelings on one side, and consider the question in cold blood as a matter of strategy: if we had been engaged in the Barents Sea, it might have incurred C-in-C's coming there and being forced to engage the Tirpitz *in the face of attacks by German aircraft. We might well have suffered a major disaster.*[26]

Hamilton was honest enough to admit to the sailors that he would be surprised if more than half of the merchant ships got through – but he urged them to put it in the context of all the convoy runs, some of which had suffered equally serious losses.

Over the coming weeks and months fights broke out in many places ashore between merchant seamen and Royal Navy sailors, the latter defending their service against accusations of cowardice in the face of the enemy.

The officers of the *Wichita* and other American warships also found it difficult at first to forgive their British allies.

In the officers' mess at Scapa, after rather too many beakers, there was much mutual recrimination and many hard words were passed. These were finally resolved in cursing the Admiralty and their inability to judge a tactical situation from a lot of pins on a board more than a thousand miles away. It was considered to be a pusillanimous defeat and a shocking error of judgement.[27]

Post Mortem

Why didn't Admiral Pound trust Tovey's judgement? The latter had warned him that scattering a convoy on the north Russian route would be 'sheer bloody murder'.[28] Or why couldn't Pound have left the final decision to Hamilton, to act as he saw fit when the *Tirpitz* and other heavy ships actually arrived?

Pound, who died from a brain tumour in October 1943, disliked delegating big decisions, but his last experience of combat had been a very fleeting part in the Battle of Jutland in 1916, as captain of the battleship *Colossus*. He knew nothing of the front line reality of war in Arctic waters.

Unwilling to trust the officer who advised him that *Tirpitz* had yet to sail, and would even then not venture far, Pound erred on the side of caution and ended up making one of the most reckless decisions in naval history.

Captain Fegen, who sacrificed his life as well as the *Jervis Bay* to save another convoy, was posthumously awarded the Victoria Cross. Not a single warship was lost in defending PQ17 and those in charge earned nothing but bitter accusations of cowardice. While acknowledging that the facts of the case provide ample circumstantial evidence for strong accusations, they are every bit as unjust as might be claims of cowardice and stupidity laid against the PQ17 merchant service sailors who surrendered seaworthy vessels, ran them aground or scuttled them to save their skins.

Tovey was angry but the man who sank the *Bismarck* was reasonably secure and his reputation intact. Rear Admiral Hamilton was the obvious scapegoat, but provided sound reasoning for taking the destroyers with him after receiving the scatter order:

> *My reason for attaching the destroyers to my force was that I assumed that* Tirpitz *was at sea and approaching the vicinity. Had this in fact been so, the addition of six destroyers to my force would have been invaluable, whereas the possibility of their being any protection to the scattering convoy was negligible in comparison.*[29]

Stalin lost some tanks, aircraft, ambulances, supplies and other equipment which he never thanked either America or Britain for trying to get through to him.

An American commentator observed twenty years later, shortly after the full facts had at last been revealed by the Admiralty to a shocked world:

> *This was one of the most badly hit convoys of the war, but for a long time there were many that were nearly as bad.*[30]

But still, Surgeon Lieutenant Ransome Wallis believed the destruction of PQ17 stained forever the reputation of the Royal Navy and reduced the service in the eyes of its own sailors.

> *The air of conscious superiority which, someone once remarked, was typical of the Royal Navy was reinforced by the knowledge that what we did had to be done and was worth doing. The deep concern felt by all ranks as to whether it was well done showed itself after PQ17.*

For HMS *London*'s ordinary sailors, PQ17 was a bitter experience that haunts many of them to this day. Sixty years on Gordon Bruty noted sombrely:

> *The story of PQ17 is embedded in my soul. We never talked about it during the war. Occasionally someone might ask you about PQ17 and you would say that you didn't know what they were talking about. This was an event that remained secret for a long time. I never met any Merchant Navy men so I didn't encounter their hostility.*

Waldie Willing, who was a leading seaman in charge of one of *London*'s 4-inch guns during the convoy, felt there could be no excuse for what happened:

At the time I felt that it was a disgrace. I think it did a lot of damage between the Merchant Navy and the Royal Navy. It was a desertion by the Royal Navy. I think the RN itself was ashamed. No one understood on the ship why we went on that wild goose chase. The truth was hard to bear when the full information came out many years later. Only now am I reading about, and finally understanding, a bit more about why we turned around and left them to be picked off.

In writing his memoirs many years later, Surgeon Lieutenant Ransome Wallis observed:

From the twenty-two merchant ships sunk in PQ17, 153 merchant seamen lost their lives. Regrettable as the loss of these merchant seamen was, it is rather astonishing that PQ17 has come to be fixed in the public mind as one of the great tragedies of the war.

He also reflected:

Sad though this was, the number of casualties was not excessive compared with previous convoys in which there were, on a number of occasions, far more men lost than this.

War is a messy, dreadful business and, in the context of the whole conflict, PQ17 was a minor disaster. The 153 mariners, most of them British, who tragically lost their lives were sacrificed needlessly. But, compared with the losses being sustained by the Royal Navy and the Merchant Navy as a whole in 1941-42 it was a minor blow. For example, during the short period 21 May 1941 to 1 June 1941, 2,000 Royal Navy sailors were killed during the evacuation of Crete. However, they were all British deaths and, like the forgotten dead of Gallipoli, they lie largely unremarked as each anniversary passes, while other nations never forget to remind the world of their own sacrifices. To this day Britain prefers to take the broader view.

Notes

1. John Clagett, *The US Navy in Action*
2. Paul Lund and Harry Ludlam, *PQ17 Convoy to Hell*.
3. ibid.
4. Robin Brodhurst, *Churchill's Anchor*.
5. Edwyn Gray, *Hitler's Battleships*.
6. J. Broome, *Make Another Signal*.
7. ibid.
8. ibid.
9. Captain J. Broome, papers relating to Convoy PQ17, National Maritime Museum.
10. ibid.
11. J. Broome, op.cit.
12. Douglas Fairbanks Jr., quoted in *Knight Errant* by Brian Connell.
13. ibid.
14. Godfrey Winn, *PQ.17 A Story of a Ship*.
15. John Clagett, op.cit.
16. Paul Lund and Harry Ludlam, op.cit.
17. ibid.
18. David Irving, *The Destruction of Convoy PQ.17*.
19. ibid.
20. Edward Reynolds, Sound Archives, Imperial War Museum.
21. Papers of D.A. Hibbit, Imperial War Museum.
22. Douglas Fairbanks Jr., quoted in *Knight Errant* by Brian Connell.
23. Paul Lund and Harry Ludlam, op.cit.
24. David Irving, op.cit.
25. ibid.
26. ibid.
27. Brian Connell, op.cit.
28. Stephen Roskill, *The Navy at War 1939-1945*.
29. Report of Proceedings of Flag Officer Commanding the First Cruiser Squadron, Public Record Office.
30. John Clagett, op.cit.

Chapter Thirteen

AN UNDESERVING SCAPEGOAT

An Admiral Leaves

In the aftermath of the PQ17 disaster, Rear Admiral Hamilton put on a brave face, yet the wolves were gathering, hungry to devour a suitable scapegoat. In a letter to his mother dated 14 July he stuck to his own security instructions and made no specific reference to PQ17, but he did give her a hint that all might not have been well:

I am not unduly perturbed about the war, I always thought this would be the critical period. I see no reason why Rommel should not suffer a major defeat in the long run. I am afraid the Russians are having a bad time, but as long as they go on killing Boches it's all that really matters.

He revealed that his cruiser squadron had just returned from escorting a Russian convoy

...where we lost a lot of good merchant ships, without doing any damage to the Tirpitz. However, that's all in the game and we have got to face up to it.

He tried to end on a cheerful note:

Don't worry about the war I am sure Hitler is finding it much more unpleasant than we are.[1]

Meanwhile, on 15 July, the Prime Minister wrote a minute to the First Sea Lord – the man who had made the fatal error – in which he said:

I was not aware until this morning that it was the Admiral of the cruisers, Hamilton, who ordered the destroyers to quit the convoy.

He also asked:

What did you think of this decision at the time? What do you think of it now?[2]

What Churchill did not know at that point was that it was Pound who had written the 'Scatter' signal and it was a certainty that the enquiry the Admiralty was conducting into the convoy disaster would not turn the spotlight on the First Sea Lord. In the end the enquiry decided not to blame anyone officially. Unofficially Hamilton was blamed, primarily for taking the close escort destroyers away.

Conscious of the bad blood between the Merchant Navy and the Royal Navy following the convoy disaster, and the fact that fights had broken out ashore, Rear Admiral Hamilton did his best to try and heal the wounds. 'I have recently organised parties of merchant seamen to visit our ships,' he told his mother in a letter from HMS *London* dated 2 August. 'They are of all nationalities, and we give them a walk around the ship, tea and a cinema, which I think they appreciate.'[3]

The intervention of illness gave the Admiralty its opportunity to temporarily remove Hamilton from command of the First Cruiser Squadron, though he would return to sea as CS1 boss and be in command until August 1943. In a letter dated 18 August 1942, Hamilton revealed to his mother that he was in a Reykjavik hospital, having had his appendix removed. At the end of the letter he cheerfully revealed an unusual present to HMS *London*:

P.S. I have made the old Chief Sick Berth Steward on board a present of my pickled appendix. This will I have no doubt be on view in the sick bay, and provide a topic of conversation in the sailors' weekly letter home – I thought it would be so selfish to keep all the fun and excitement to myself.[4]

An Artificer Joins

Young sailor Ron Wood joined HMS *London* at Scapa Flow in September 1942, along with another Engineering Artificer 5th Class called George Sindall. Through Wood's experiences we get an insight into how grim it was to live aboard *London* that autumn as she made up part of the cruiser screen for convoy PQ18.

> *...as dusk was falling a naval pinnace arrived and George and I fell aboard. Apart from the cox'n checking our names and numbers, no one spoke as we sped across the water that had now started to get rough. Apart from the rain we now had to turn our heads away from the cold spray that tried to slap our faces... The pinnace made its way through the many vessels at anchor, the vast grey battleships ominous shadows against the grey sky – shadows that appeared and disappeared, with no sign of life, no warmth of humanity. This was my first sight of the fleet at war. I looked at George and knew he was sharing my feeling of despair. The* London *appeared through the gloom, 10,000 tons of steel, cold and colourless. As the pinace drew closer we could hear the low hum of the mass of machinery that pulsates within the armour plating of a cruiser.*[5]

He found that life aboard a warship was an assault on all the senses:

> *A cruiser at sea in wartime is a complexity of sounds and smells. It bustles with activity, yet is still and silent. It has light and heat, yet is cold and humid, like a tomb. It has the freedom and space of the upper deck, and then there are the claustrophobic locked compartments during action stations. The companionways that run throughout the ship, from stem to stern at all deck levels, are fitted with heavy watertight doors set at regular intervals. Every compartment has a hatch or door that can be securely fastened, usually with six dog clamps. Each watertight door is fitted with a steel plate welded across the door opening, the plate being about two feet six inches high. Its purpose is to hold back water while the door is being closed, in the event of flooding caused by enemy action. During normal cruising the watertight doors are left open to allow the crew to go about their business, the steel plates across the door openings however, make the journey through the ship rather like a hurdle race.*

The call to action stations was sounded by a Royal Marine bugler via the ship's loud speaker system. This had the immediate effect of filling the companionways and ladders with struggling humanity, always, it seemed, intent on going in opposite directions; every man had an action station to report to in the shortest possible time.

> *The companionways are pandemonium, with watertight doors being opened and closed in urgent desperation, as men fight past each other to get to their posts. The noise of the heavy doors banging shut, and the dog clamps being knocked up tight add to the confusion, bruised and split fingers were not uncommon... the chaos is short lived however, and an ominous silence settles over the ship. It is as though all the beating hearts of the now closed-up crew are merged with the whirring machinery that is the beating heart of the great ship, as she ploughs through the grey Arctic Ocean.*

Having had the need for cleanliness and orderliness aboard a naval ship drummed into him, Ron Wood was shocked by the disorder and decay in *London*.

> *Normal routine like cleaning ship was a non-event. As the days passed so the ship became more and more untidy. Spilt food on the mess decks remained there... cocoa and sugar were all over the decks. I can hear the sound of granulated sugar crunching under my feet to this day. The seamen's heads, or toilets, were in the forecastle and I remember seeing a line of white porcelain lavatory pans disgorging their contents in fountains every time the great ship put her bows down into the deep arctic troughs. Handrails on the vertical ladders throughout the ship*

HMS *London* 1942-43. *G. Bramley Collection.*

were rusty. Even the main throttle valves in the engine room had not escaped the cold, moisture-laden air. Seasickness was not uncommon and on one occasion I saw the ship's cat being sick in a fire bucket. The cat was more considerate than some members of the ship's company.

But, sometimes the young sailor's peers were inspiring:

A man I greatly admired was the stoker Petty Officer of the watch, a giant of a man with many years of naval service, gentle and unassuming. He would arrive in the boiler room always in clean and ironed overalls, shaved and not a hair on his head out of place. He would position himself with his back to a boiler, feet astride and hands clasped behind his back. First his right knee would slightly bend, and then his left, taking in the rhythm of the roll of the ship, only his eyes moving to observe the pressure gauges and the performance of the stokers at the furnace fronts. The quiet confidence of this man stemmed from long and hard experience, a four-hour watch in the boiler room on the Russian run was like shelling peas to him. At the end of his watch he left the boiler room as fresh and alert as when he had entered.

The forced-draught fans that served the boiler rooms, maintaining a pressure above atmosphere, were situated way above the boilers. These fans sucked great volumes of Arctic air down into the bowels of the ship. On one occasion I witnessed the strange sight of snowflakes falling in the boiler room. As they silently fell upon the stoker Petty Officer I could hardly believe my eyes. A reassuring smile on his face told me he had seen it all before.

But, such signs of warmth and humanity amid the cold grey steel of a cruiser at war were few and far between. After nearly two years enduring Arctic conditions, and with the recent demoralizing experience of PQ17, Ron Wood found the *London*'s crew were all but drained of kindness and compassion. Witnessing a refuelling at sea provided a chilling insight into how close terror lurked beneath the numb exterior.

From out of the gloom appeared HMS Eskimo, a Tribal Class destroyer. As she came in to take up position, we looked down on her decks and the men waiting to receive the oil hose. There were no friendly greetings between the crews of the two ships, men of the convoys had long since abandoned friendly trivialities – war had sapped the warmth of humanity and left them only intent on survival. Although the speed of the two ships was only eight knots, the gap between them was a torrent of white icy water. The hose had been connected and pumping had been in operation for about five minutes, when suddenly the gap between the two ships opened up. The oil hose parted and hot fuel oil spewed into the sea and over the decks of the ships, as the hose whipped around like a wounded python. One of the hose guidelines on the destroyer

parted and caught around the leg of a Chief PO and dragged him across the deck and over the side of the ship. As he went over he just managed to grab a deck stanchion, his terror-stricken eyes looking up at us. A seaman dashed out of the galley on the upper deck, with a meat cleaver, and cut away the rope and with help hauled the chief back on board. The destroyer made hard to starboard and disappeared into the gloom.

Convoy PQ18 was a hard fought run, but its escort force was substantial and stuck with it all the way. The Home Fleet flagship was still *Duke of York* but the new King George V battleship *Anson* was now in commission and the two battleships and escorts provided distant cover, while closer at hand was a five-strong cruiser force led by HMS *London*.

The *London* had departed Hvalfjord on 14 September in company with *Norfolk*, *Sussex*, *Cumberland* and *Sheffield*, together with destroyers. Another cruiser, the *Scylla*, was with the convoy plus the usual gaggle of corvettes, AA ships, minesweepers, armed trawlers and two submarines. But the most significant evidence of greater, and more determined, protection was the

A boxing match on the *London* in the Hvalfjord.
W. Willing Collection.

One of *London*'s young sailors poses on her upper deck, with *Duke of York* visible ahead. *W. Willing Collection.*

HMS *London* 'takes it green' over her bows in the Arctic. *G. Bramley Collection.*

assignment of a new escort carrier, HMS *Avenger*, carrying Hurricane fighters and Swordfish torpedo bombers, together with a sixteen-strong destroyer strike force.

Further air cover and assistance was to be provided by RAF Catalinas flying from an airfield near Archangel and there was also supposed to be support from RAF bombers. But most of them crash landed on arrival in Russia and one was even shot down by Soviet AA guns.

While there was no way the well-informed Germans would risk sending out their surface battle group to face such a strong escort, they did pitch in torpedo bombers, Stukas, Condors and U-boats, with a six day running battle developing as the convoy progressed to Archangel.

The British cruiser force patrolled between Bear Island and Spitzbergen, coming to within 700 miles of the North Pole. A joker aboard *London* thought it might add a nice touch of irony if a record of Deanna Durbin singing *No Place Like Home* was broadcast over the ship's loud speaker system.

PQ18 suffered ten out of forty merchant vessels sunk while QP14, coming the other way, lost three out of fifteen supply ships. Two British warships were also lost. The Germans paid a high price, losing forty aircraft and three U-boats.

After PQ18 and QP14 it was decided to give the convoys a rest, particularly with resources needed for the forthcoming Torch landings in North Africa. Some ships did make the trip to Russia on their own, without escorts, and half of those setting out reached their destination.

The *London* got back to Iceland in late September, conducted a couple of patrols in the Denmark Strait then went down to the Clyde. After escorting a trans-Atlantic convoy she returned to Iceland. Shore leave in Scotland had restored a measure of black humour, with some of *London*'s officers trying to earn their tickets out of the Arctic by pretending to have a 'mythical dog'.[6] It 'lived' in the wardroom where it even had its own lead and collar hung on a hook, and a water bowl in a corner.

Engineering Artificer Ron Wood had already made his escape, leaving *London* immediately after PQ18.

Just when it seemed to George and I that we were destined to endure the misery of the Northern Patrol for the duration of the war, a miracle happened. On reaching Iceland we were told to pack our bags and hammocks, we were on draft back to Chatham.

Telegraphist D.A. Hibbit had also escaped, ending up at a training establishment in Plymouth that had been a nudist colony before the war. It was a bit basic, but it was better than being stuck on HMS *London* in the Arctic.

Tirpitz Remains in her Lair

Before the Allies imposed a temporary halt in the Russian convoys, the twenty-eight vessels of QP 15 had been escorted back from Murmansk to Iceland, with the *London* and the *Suffolk* providing a cruiser screen from Bear Island to Iceland. There was little opposition from the Germans, with QP15 losing only two merchant vessels to U-boat attack. The Allied invasion of North Africa in November 1942 had the benefit of drawing away German aircraft and submarines. With the onset of twenty-four hour darkness in December, the Arctic convoys were resumed with the prefix JW in place of PQ and RA in place of QP. The *Tirpitz* remained in her lair, but the *Scheer* was in refit. The *Hipper*, light cruiser *Nürnberg* and the *Lützow* were, however, still in Norwegian waters. Hitler put heavy pressure on the naval high command to mount a major operation against the resumed Arctic convoys, but the same restrictions were still in place. Yet even the mere presence of the *Tirpitz* and the other ships constituted a huge threat. *London's* sailors groaned at the increasing frequency of Patrol White and it was becoming plain that HMS *London* would not last much longer without further major repairs. This offered a glimmer of hope.

HMS *London*'s Y turret crew in their winter gear. *G. Bramley Collection.*

The autumn and early winter gales in the Denmark Strait were now playing havoc with HMS London which was cracking badly. By November buzzes about a refit were rife and the main speculation was about 'where and when'. In early December we returned to Scapa Flow and it then became known that we were to go down to the Tyne... .[7]

With her paying off pennant trailing behind her, *London* sailed up the Tyne to Middle Dock at North Shields just before Christmas 1942. Surgeon Lieutenant Ransome Wallis found himself one of the last officers to get a new posting and while some might have disliked their service in HMS *London*, the naval doctor regarded her with some affection, calling her a 'fine sea boat'.

The refit lasted five months, and involved considerable strengthening, 'after which she was satisfactory as any ageing, hard-worked ship'.[8] A considerable effort was made to repair the oil tanks and *London* was provided with special equipment to remove water from them. The cruiser's aircraft launch gear was removed and her AA fit was boosted, with seven additional 20-mm cannons (eight single 20-mm cannons had been fitted during the previous refit, in place of her machine guns).

By 4 May 1943 HMS *London* was back at Scapa Flow and on sea trials prior to resuming convoy escort work and patrols in the Denmark Strait. Joining her at Scapa was sixteen-year-old Signalboy William West who found the *London* a home-from-home.

As a cockney I was chuffed to bits that my first ship was named after my home city. The fact that they named the various sailors' messes and other places after boroughs of London was even better. The aircraft hangar, which was used for film shows, was named 'Highgate' while my mess was called 'Woolwich'.

Espionage off Spitzbergen

The *London* played the role of the *Tirpitz* during a training exercise in which the Home Fleet carrier HMS *Illustrious* flew off a 'strike force' of Wildcat fighters and Barracuda torpedo bombers to find and 'attack' her.

In July she took part in two diversionary moves that aimed to distract attention away from the Allied landings in Sicily by reinforcing Hitler's fear about an invasion of Norway. Around the same time *London* took part in Operation Holder, the conveying of supplies, personnel and mail to Royal Navy bases in north Russia.

Yet more Patrol Whites followed in September and October. Quarters Rating Bob Houghton, who helped transfer the Moscow mission members from HMS *Harrier* to the cruiser two years earlier, had joined *London* while she was still in refit. He recalled that an American cruiser suffered serious structural failure in the Denmark Strait during one of the October patrols.

It was a miserable boring period of cold and terrible weather and we were in the company of the USS Augusta but her upper deck split open and she had to return to port for repairs.

Signalboy West was frequently exposed to the elements, as he worked on the flagdeck of the cruiser, running up signal flags or communicating with other ships via light signals.

You were well protected, in a special padded suit with an oil skin exterior, but you had to be careful you didn't touch the metal signal lanterns with your bare hands or you would leave your skin on them.

In October 1943 HMS *London* was sent on an espionage mission to waters around the island of Spitzbergen, far to the north of Iceland, where it was suspected the Germans were

Ice slush surrounds *London.* W. Willing Collection.

maintaining a secret radio intercept station. The cruiser sailed around the island, her own communicators waiting for tell-tale signals from the Germans, but they kept off air. The fact that a Luftwaffe scout plane kept doggedly overhead monitoring the *London*'s movements was probably enough of a tip-off to keep quiet.

Late in the same month the cruiser helped provide cover for the convoying of five US-built minesweepers and six metal hulled motor launches to Murmansk. Operation FR started from Akureyri in Iceland and, shortly before it departed, the Soviet Navy sailors who would crew the convoy vessels were entertained on board the *London*. This was a rare instance of Anglo-Russian fraternization for, at British naval bases in the USSR, contact between Royal Navy sailors and the local population or naval personnel was banned.

Able Seaman Gunner Benny Goodman found the Russians who came aboard *London* to be friendly enough.

> *We gave them afternoon tea in our aircraft hangar and they were particularly excited by the white bread used to make the sandwiches, as they had never seen it at home. Later they went down to the ship's canteen shop and virtually cleared us out of cigarettes and anything else that took their fancy. One of them was a woman who, apparently, had been a sniper during the siege of Leningrad where she killed over 100 Germans. It may have been a bit of leg-pulling Soviet propaganda... but there again she looked like she could have done it.*

For *London*, service in the Arctic was drawing to a close as she was about to receive orders for another diplomatic mission, this time carrying members of the Prime Minister's staff to Egypt. As she sailed south to Plymouth in early November to pick up the mission members, the *London*'s crew filed away their memories of war in the Arctic. Graham Bramley had seen plenty of German airmen left to die in *London*'s wake after being shot down, but some of them were picked up and one in particular had been a revelation:

> *On one convoy this German pilot knew his plane was badly hit and wouldn't make it back to base, so he deliberately crashed it beside us and we pulled him aboard. He was lucky as he*

only had frostbite on his hand. I was sent down to remove a ring from his finger His hand was like a balloon and the ring was cutting into him, so I used a jeweller's hack saw. He could speak English and, as he was the first German I had been in close company with during the war, I asked him if he could tell me what all the fighting was about. He told me he'd much sooner be at home than trying to sink us. He was quite a nice chap and I was surprised at how well I got on with him, for someone who was 'the enemy'.

The strange remoteness of attacks on nearby merchant ships also stuck in Graham Bramley's memory:

You might be clearing ice on the upper deck and see a merchant ship being hit. You saw the explosion... the smoke... the ship becoming a funeral pyre... then you carried on working. Once people went in the water they stood no chance and thinking about it would have made you go insane. I guess you numbed yourself against it by hiding in the daily routine on your own ship.

Engineering Artificer Ted Huke had managed to block out the horror of what he might one day have to do to his shipmates, with some studying.

You were on your own on the end of a phone, waiting for the order to flood the magazine, drowning everyone in it to save the ship, which you hoped would never come. To occupy my mind I did a bit of swotting with some technical manuals. After six months it was decided that they were wearing the crew out so they allowed us camp beds down there.

Gunnery rating Waldie Willing was one of those who would have been drowned if the worst had ever happened:

I went aboard London *as a Boy Seaman and then was an Ordinary Seaman and went into the gunnery branch. I started off in the 8-inch magazine. You loaded all the cartridges of cordite onto a roller system to go into the hoist which took them up to the turret. I didn't like it right in the bottom of the ship knowing there were people above us who would have been ordered to flood the*

Two sketches by *London's* **Ron Wood depict the grim reality of being closed-up at Action Stations in the Arctic, waiting for the order to flood compartments.** *R. Wood.*

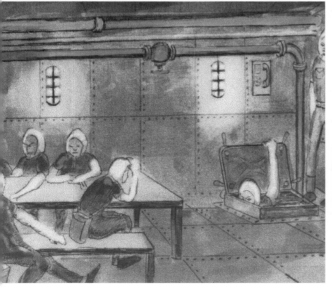

HMS *London*'s ship's company in 1943, prior to the ship leaving for the Far East. *W. Willing Collection.*

magazine if we were hit badly. I was glad that I passed my exams and I got my gunnery qualifications allowing me to work my way up the gun. You went into the shell room after the magazine and ended up in the turret itself. It was much better than being in the magazine I can assure you. Being on the Arctic convoys was the worst job in the war. If it was bad for us in a cruiser like the London*, it was even worse for merchantmen that couldn't really fight back. One of my duties was watch-keeping and goodness me what seas!* London *often put her nose right under. Often people were not allowed on the upper deck because it was so dangerous. Inside the ship it could be a bit nasty – but it wasn't too bad really. There might have been a slight lack of house-keeping due to the constant rolling of the ship. It was very soul destroying up there in the Arctic, but we survived.*

Notes

1 Papers of Admiral Sir Louis Henry Keppel Hamilton, National Maritime Museum.
2 David Irving, *The Destruction of Convoy PQ.17.*
3 Papers of Admiral Sir Louis Henry Keppel Hamilton, National Maritime Museum.
4 ibid.
5 Papers of R. Wood, Imperial War Museum.
6 R. Ransome Wallis, *Two Red Stripes*.
7 ibid.
8 D.K. Brown, *Nelson to Vanguard*.

Chapter Fourteen

SURRENDER AT SABANG

Carrying the Stalingrad Sword

Arriving at Plymouth on 10 November, HMS *London* embarked members of the Prime Minister's staff, while Churchill himself sailed to Egypt in the battlecruiser HMS *Renown*.

The tide of the war had by this time turned decisively against the Axis powers. The Americans had inflicted a decisive defeat on the Japanese at Guadalcanal, while North Africa had been cleansed of German and Italian troops and the Allies were now rolling up Italy after taking Sicily. On the Eastern Front the Russians had delivered massive blows against the Germans at Stalingrad and Kursk.

In Cairo, Churchill, Roosevelt and the Chinese Nationalist leader Chiang Kai-shek discussed a strategy for defeating Japan. The Americans were to continue with their advance via the island-hopping campaign in the central Pacific and through the Philippines, while the British liberated Burma, Malaya and the East Indies. The Chinese agreed to carry on supporting Allied offensive operations in Burma if possible.

The Cairo conference was a precursor to the first 'Big Three' meeting in Tehran where Churchill, Roosevelt and Stalin agreed the final strategy for defeating Germany. This included an invasion of north-west Europe and a simultaneous Soviet offensive to eject the Germans from the Ukraine. Russia promised to enter the war against Japan after defeating Hitler. At the conference Churchill presented the Soviet leader with the famed 'Stalingrad Sword', a gift from King George VI to the people of Stalingrad that celebrated the great victory over the Germans. It had been carried from Plymouth to Egypt by HMS *London* before being put on the plane with the British delegation. During its time in the cruiser, shifts of four Royal Marines stood guard over the casket in which it was contained.

HMS *London* at high speed astern of the battlecruiser HMS *Renown*, during their passage to Cairo carrying Winston Churchill and his staff. *Bill Jacobs Collection.*

The *London* departing Alexandria in November 1943 following 'The Big Three' conference in Tehran. *Bill Jacobs Collection.*

The *London* stayed at Alexandria during the two conferences and in early December set out on the journey home carrying VIP passengers including Geraldo's famous orchestra. It provided a soundtrack for the return, including a new tune, composed in honour of the cruiser, called *The March of the London*.

Arriving at Gibraltar on 13 December, *London* left for the UK four days later, enduring rough seas on the way. On 21 December she arrived in the Clyde, disembarked her passengers and two days later sailed for Rosyth where she was to undergo another refit. More AA guns were fitted, with twin 20-mm weapons clustered around one of the funnels and a new 40-mm cannon on either side of the bridge.

After reaching the tropics HMS *London*'s ordnance artificer senior rates and their commanding officer pose for a portrait in front of the forward 8-inch guns. Graham Bramley is far right, front row. *G. Bramley Collection.*

By early February 1944, she was back on the Clyde embarking stores, mail and new drafts of sailors. Extra blankets were distributed to the cruiser's sailors, causing many of them to sink into despair – it was the traditional sign for *London* heading back north and another dose of Patrol White. However, it was merely a ruse to prevent careless talk from giving away her true destination for, when *London* put to sea in company with the cruiser HMS *Cumberland*, she was heading for the Far East, not the Arctic. On 13 March, *London* arrived at Colombo, one of the two major British naval bases in Ceylon and, four days later, transferred to the other, Trincomalee.

Striking Sumatra

After intensive fleet exercises, *London* sailed as part of the cruiser screen for the Eastern Fleet's first offensive in two years.

A twenty-eight vessel task force had been gathered for air strikes against Sabang in Sumatra. Led by the fleet flagship, the battleship HMS *Queen Elizabeth*, the task force also included the carrier *Illustrious*, *Queen Elizabeth*'s sister vessel, *Valiant*, and the French battleship *Richelieu*. In lieu of more British carriers arriving in the Far East, the US Navy had agreed to attach the USS *Saratoga* to the Eastern Fleet for the operation.

On 19 April, the task force attacked the island of Sabang with no serious opposition from the Japanese. Aside from damaging defences and logistics infrastructure (such as airfields, oil storage facilities and power stations) in an area that might soon be invaded by British troops, the Sabang attack also distracted the Japanese from an American offensive in New Guinea.

On 30 April the Anglo-American task force set sail again, this time heading for Exmouth Gulf, in western Australia, where the ships anchored and proceeded to take on oil and ammunition. By 17 May the task force was off Surabaya, in the Dutch East Indies, with *Illustrious* and *Saratoga* again launching air strikes and a bombardment, with *London* providing close escort for the carriers. The Japanese opposition during the day was light

The raid on Surabaya. *Ballantyne Collection.*

Teenage sailor John Reekie. *J. Reekie Collection.*

with just two aircraft lost, one on take-off from its carrier and the other to AA fire. As the task force withdrew that night, Japanese torpedo bombers made a lunge towards *London* and the *Saratoga*. Able Seaman Gunner John Reekie got a good view of the attack as he was manning an Oerlikon cannon on the port side aft.

It was between 10 p.m. and 11 p.m. and we could hear the Japanese aircraft coming at us and they must have dropped their torpedoes a fair bit out. I saw the phosphorescent wake of one of them coming straight at us. The 4-inch guns opened up on that bearing... blinding and deafening me with their smoke and flame as the London *turned sharply. The* Saratoga, *which was to starboard also turned to starboard pretty sharply. If the* London *had not turned I am sure it would have hit her right in the stern.*

The task force dispersed on 31 May, *London* arriving at Colombo on 1 June. After a two week visit to Trincomalee, she headed back to Australia, calling in at Fremantle and, after giving her sailors some shore leave, escorted a troopship to Ceylon.

On 12 September, after a period of fleet exercises, *London* arrived at Bombay where she entered dry dock for bottom cleaning and essential maintenance. Thirteen days later *London* left Bombay, carried out some gunnery

One of *London*'s sailors is held down during a crossing-the-line ceremony in 1944. *G. Bramley Collection.*

HMS *London* making smoke furiously during the bombardment of Car Nicobar.
Illustration by Dennis Andrews, taken from a photograph by Alexander A. Whitcombe who was serving in the cruiser HMS Cumberland.

practice, and headed for Ceylon where she was to join a bombardment mission in mid-October against Car Nicobar in the Nicobar Islands, north of Sumatra. The task force was spearheaded by the fleet carrier *Indomitable* and the *Renown*. As the ships began their lethal parade past the island, in line astern, *London* was at their head, with Able Seaman Reekie getting a grandstand view of the action from the Air Defence Position, where he helped coordinate the anti-aircraft guns.

> *I think* Renown *might have been right behind us and she made a marvellous sight. All the ships' big guns were firing full broadsides, including our own 8-inchers so it made an overpowering sound. The shock waves and cordite washed over you. The ships went up and down firing first on the port side and then the starboard after turning around. You could see the flames and smoke as they ripped up the greenery. Guy Fawkes night is nothing compared with the bombardment of Car Nicobar.*

Bob Houghton, recently promoted to Petty Officer, was in one of *London*'s 8-inch turrets and found the bombardment provided some welcome relief from boring routine.

> *My turret fired 52 rounds on 17 October and 10 more at midnight on 18 October and by the following day we were back at Trinco. The next eight weeks were taken up by the usual harbour routine, exercises, etc. To be honest I found my time on* London, *having served on small ships like the* Harrier *in Archangel, somewhat of an anti-climax.*

Birth of the East Indies Fleet

By the late summer of 1944, the Eastern Fleet had become the most powerful naval force that Britain had. Aside from the carriers *Illustrious* and *Victorious*, the battleships *Queen Elizabeth*, *Valiant* and *Howe*, it also included various escort carriers, the *Renown*, plus cruisers and destroyers. The Fleet Air Arm was now in the vanguard of offensive operations, as it finally had the aircraft it needed, including American-made Avenger dive-bombers, Corsair and Hellcat fighters, together with British-made Firefly and Seafire fighters.

At the Anglo-American conference in Quebec in September, Britain had offered to support the American drive through the Pacific by forming a new fleet. The British Pacific Fleet was the result, taking away the fleet carriers and the most modern battleships, destroyers and cruisers. The residual Eastern Fleet was renamed the East Indies Fleet in November 1944 and tasked with supporting the thrust through Burma towards Malaya and into the East Indies. The *Queen Elizabeth* was flagship and among other ships retained were *Renown*, several escort carriers, eight cruisers, (including HMS *London*) and two dozen destroyers. The French battleship *Richelieu* also stayed.

On 8 April 1945 the East Indies Fleet's Task Force 63 departed Trincomalee for another bombardment mission. The warships left in two groups, with *London* in Group One alongside *Queen Elizabeth, Richelieu* and the destroyers *Verulam, Vigilant* and *Saumarez*. Group Two included the escort carriers *Khedive* and *Emperor*, the cruiser *Cumberland* and the destroyers *Virago* and *Venus*.

The primary objective at that moment for South East Asia Command was to open up the Malacca Strait, paving the way for an invasion of Malaya that October. The battleships and HMS *London* were returning to bombard Sabang, because the island guards the northern end of the Strait, while *Saumarez, Verulam* and *Vigilant* shelled other targets. The only dangerous moment came when a Japanese aircraft suddenly lunged out of the clouds and dropped a bomb among the ships.

On 13 April, the *London* set sail for Simonstown naval base in South Africa where she was dry-docked for a refit. Bored though he was by his service in the *London*, Bob Houghton was fascinated by some 'new' weapons the cruiser received:

> These ex-Army 40mm Bofors AA guns were fitted wherever space could be found, no doubt in readiness for our eventual transfer to the Pacific Fleet where we would face kamikazes.

In contrast to the years in the Arctic, *London*'s sailors were now at risk from overdosing on Vitamin C. Charles Cox recalled:

> As cooks we were always glad to be in port where we could get access to fresh produce, but in South Africa we could get as much fruit as we wanted, which was somewhat different from the Arctic where we didn't see any.

Able Seaman Reekie was allowed ashore one hot night and he decided to take his mind off what dangers might lie ahead by taking the train up to Cape Town.

> When I got off at the other end, everybody was going bonkers and this woman ran up to me and gave me a big hug shouting 'the war's over!' Turned out the Germans had finally surrendered. I wouldn't have thought South Africans would get too excited, it being so far away from them. But they did and it was a heck of a celebration.

In the meantime Petty Officer Houghton was coping with naive young officers during shore patrol duties in Simonstown.

> I drew the short straw, so to speak, on VE Day. I was PO of the Canteen Patrol with a very young and green Sub Lieutenant who thought he could curtail the joyous celebration by closing the canteen at the proper time.

Petty Officer Houghton advised the officer that it might cause a riot and so *London*'s sailors got a little bit more time to celebrate.

Japan Capitulates

In the spring and early summer of 1945, Japanese resistance was fiercer than ever, with more than 12,000 Allied servicemen killed during fighting for Okinawa. Twenty-six naval vessels were sunk by kamikaze attacks and another 168 badly damaged. It was estimated that Allied forces would suffer a million casualties when they invaded the Japanese home islands. The

A launch carrying the Japanese surrender delegation passes the British destroyer *Rocket*, which was providing escort for the *London*. *Benny Goodman Collection.*

Allies decided that they had to cow the Japanese with a terrible new weapon constructed by American and British scientists – the atom bomb.

On 6 August the first bomb was dropped on Hiroshima killing 80,000 people. The second was detonated over Nagasaki three days later, killing more than 40,000. Even then many Japanese military commanders and politicians wanted to carry on fighting, but Emperor Hirohito agreed to an unconditional surrender. Across the Far East naval vessels prepared to take the surrender of individual Japanese garrisons. The Royal Navy's official account of HMS *London*'s war service records her part in this:

> On 28th August, flying the broad pendant of Commodore A.L. Poland, she arrived at Sabang, and it was on board her that Japanese officers accepted the arrangements for occupation to become effective as soon as the instrument of surrender had been signed at Tokyo.[1]

In fact the *London* arrived on 27 August and Benny Goodman was one of those looking on as *London*'s triumphal entry into the main harbour at Sabang hit a flat note.

> We sailed in with flags flying, ship's company on deck at attention in best whites and the Royal Marine band going full blast with 'Hearts of Oak'. Then visual signals to our ship from the Japanese shore station shook things up. It appeared that due to a mistake we had arrived a day early. Ashore we could see that all Japanese gun positions were fully manned by steel helmeted troops. The ship's tannoy ordered all upper decks, plus bridge surrounds to take shelter below. A Japanese lieutenant arrived alongside in an armoured launch. Coming aboard, and speaking in perfect English, this young officer indicated that his side was not yet surrendered and was, until the following day, still at war. Things were smoothed over and the only view we were allowed of Sabang until the following day was through portholes.

In the morning the Japanese lieutenant came back aboard *London*, to discuss terms of reference for the official surrender ceremony that would take place in six days time. As he walked across the cruiser's quarterdeck escorted by one of *London*'s young officers, the Japanese lieutenant revealed that he had been educated at Cambridge University before the war. His counterpart was delighted to be able to reply that he had attended Oxford.

On 3 September, the day after the Japanese signed the official surrender document aboard the American battleship USS *Missouri* in Tokyo Bay, Commodore Poland accepted the formal capitulation of the Japanese garrison on Sabang from Vice Admiral Hirose of the Imperial Japanese Navy. Graham Bramley, who had been promoted to Chief Petty Officer in 1944, again found his expectation of the enemy confounded by reality:

When the boat containing the Japanese party came alongside the London *we were surprised to see these six foot tall people, which were not what we imagined at all from all the propaganda we had been fed. But the Japanese on Sabang were a hand-picked regiment.*

Then someone decided to play a prank that threatened to disrupt the surrender timetable:

Once the Japanese officers had gone to discuss surrender terms, their boat remained sat alongside us, with a rather tempting Japanese Imperial Navy pennant flying from its stern. It was only a matter of moments before this sailor nips down and takes the flag as a souvenir, much to the consternation of the Japanese chaps crewing the boat. When he returned to his boat, the most senior Japanese officer was not amused. Apparently the souvenir flag had been presented to the Japanese unit by the Emperor himself. He told our captain that, unless the flag was returned, and honour was therefore restored, he was planning to go and commit Hari Kari... ritual suicide... outside the captain's cabin. Seeing that the Japanese officer meant business, the Captain got on the speaker system and explained the situation to the crew. He ended by saying that the sailor with the flag could keep it, but he would be the one tasked with cleaning the mess up. The flag was produced shortly afterwards and returned to the Japanese.

The Japanese armoured launch arrives alongside London's **starboard gangway.**
Benny Goodman Collection.

The Japanese delegation passes a Royal Marine honour guard on HMS London's **quarterdeck.**
Benny Goodman Collection.

Surrender signed, the 5,000-strong Japanese garrison boarded a collection of transport ships and sailed for Singapore where they would go into internment.

Two boat loads of *London*'s Royal Marines were sent ashore, swiftly followed by a detachment of sailors, including a colour party, and the ship's band. Having disarmed members of the local population, who had decided to try on some left behind Japanese uniforms and equip themselves with abandoned rifles, the British contingent held a flag raising ceremony on behalf of the colonial ruler of Sabang. 'I believe the island was the first piece of colonial territory in the Dutch East Indies to be taken back from the Japanese', recalled Benny Goodman, who was one of the colour party sailors. 'We ran up the Dutch flag and the White Ensign side by side.'

There was a massive amount of enemy ordnance to be cleared up on the island, including thousands of shells that were temporarily stored in caves out of harm's way, and hundreds of guns of all shapes and sizes. Graham Bramley was nearly blown up:

> I was put ashore to check all these anti-aircraft guns in case some clown was tempted to fire them. I took a marine with me. This marine says: 'What's all this plasticine?' I looked over and saw he was mucking about with some of it. I shouted: 'Put it down! That's plastic explosive' ...or words to that effect. The Japanese used to melt it to put in their shells and that's why there was plenty of it lying around to tempt unsuspecting Royal Marines.

Leading Cook Charles Cox was another member of the landing party.

> I set up a field kitchen for our people, but we also ended up feeding the locals. There was no food left – the Japanese had eaten everything. I did find a room stacked high with empty wine bottles, but that was about it. We found some rice, which I boiled up to feed the locals, and the ship sent food to us.

Other sailors were engaged in smashing up the many bottles of saki the Japanese had left behind, to make sure it didn't fall into the hands of the local population who might become drunk and start a riot.

Arriving at Singapore in mid-September, HMS *London*'s sailors were shocked at the condition of British servicemen who had just been released from the notorious Changi prison. Benny Goodman's mess invited some of them down for a cup of tea.

> They were living skeletons and we were told not to feed them, because their stomachs wouldn't be able to take it and they would die. I sat opposite this bloke who saw a tin of Brylcream on our mess table and, before anyone could stop him, he seized it, scooped it out and swallowed it down. He hadn't had any fat for years. We were really choked up.

HMS *London* in Far Eastern waters, 1945. *P. Seaborn Collection.*

Some of *London*'s gunners were asked to go ashore as members of naval shore patrols to enforce law and order.

On returning from this duty some of Benny Goodman's shipmates told him about a heartbreaking discovery:

> They went into some rooms in Changi Prison where British prisoners were usually kept the night before the Japanese executed them. On the walls they found some last messages the prisoners had scrawled, sending love to their families... their last goodbyes to their parents or wives and children. Some officer said it would be bad for morale if the people back home found out about these messages, so they were covered over with whitewash. These guys had left their personal details so someone would be able to track down their relatives back home. Somebody should at least have photographed the messages, so their families could see them.

On 14 October, at Colombo, *London* embarked around 400 passengers and 200 bags of mail and the following morning set sail for home, arriving at Sheerness three weeks later. By 9 November, HMS *London* was alongside at Chatham but Petty Officer Houghton could not go home just yet:

> As the London *was going to be used as a troopship, carrying people to and from the Far East and Australia, we who were due for demobilisation stayed on board while the remainder had their leave. On 23 November I left* London *and three days later I was demobbed and became a civvy once more.*

Graham Bramley recalled that Britain's war weary capital city did not celebrate the safe return of its namesake ship with an official reception in her honour:

> The London *was a very lucky ship – we went right through the war and were never hit but there was no welcome for us when we came home. They just sent us straight back out to the Far East on trooping duties.*

Gordon Bruty, who had been with the cruiser throughout her five long years of war and was, by 1945, an Engine Room Artificer 1st Class, also regarded her as being cruelly unsung:

> She served and endured, coming through it all without a scratch and while there was no great fanfare when she got back, the London *was still a great ship.*

Others might have been glad to escape her during the Second World War, but Benny Goodman regarded the *London* as the best cruiser in the fleet.

> I know that some people might have seen things differently, depending on when and where they served in her during the war, but I thought she was a very happy ship. Some might not have liked the food but I thought she had good cooks and she was full of fighting spirit too even though she was not in any of the big battles.

The *London* might have come through the Second World War unscathed and unnoticed, but ahead lay her sternest test under fire and it would make headlines around the world.

Note

1. *HMS* London *Summary of Service*, Royal Navy Historical Branch.

Chapter Fifteen

THE FINAL DAYS OF EMPIRE

Retired and Reprieved

The *London* left for her first trooping trip (UK to Colombo) on 26 November 1945 and set sail on her second (UK to Sydney) on 19 January 1946. Returning to Britain, she entered Devonport Dockyard for a short refit before embarking on her third trooping trip that May, carrying troops to Singapore.

Returning to Plymouth in late June, HMS *London* was taken out of service. This was the official conclusion of her 1941-46 commission and her flag was laid up in the Tower of London that September. It was estimated that *London* had steamed 257,150 miles during the commission, with the lowest annual mileage being 28,800 in 1943 and the highest 50, 921 in 1945. She had carried 4,075 passengers since October 1945.

More than half a century later Junior Rating Peter Seaborn recalled some 'highlights' of his time serving in *London*:

> After leaving Malta one time in 'heavy weather' I recall we had to return to land casualties caused by the extreme conditions. And when we called at Port Said it could be a bit risky getting detailed as a 'buoy jumper' – i.e., securing the ship to a buoy – as you were often threatened by knife-wielding Egyptians. I suppose they thought we were stealing their work. We ended up being protected by Royal Marines with rifles. I remember food was not always very enticing. We used to examine every piece of bread, as it was normally impregnated with weevils, mainly dead. In fact I used to avoid the main courses and survived on the sweet. I remember we were so

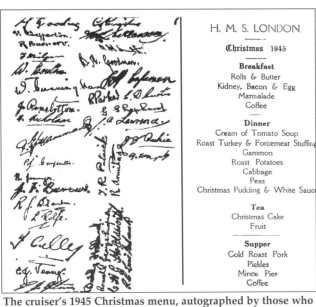

H. M. S. LONDON.

Christmas 1945

Breakfast
Rolls & Butter
Kidney, Bacon & Egg
Marmalade
Coffee

Dinner
Cream of Tomato Soup
Roast Turkey & Forcemeat Stuffing
Gammon
Roast Potatoes
Cabbage
Peas
Christmas Pudding & White Sauce

Tea
Christmas Cake
Fruit

Supper
Cold Roast Pork
Pickles
Mince Pies
Coffee

The cruiser's 1945 Christmas menu, autographed by those who enjoyed it. *J. Reekie Collection.*

Enjoying a good run ashore in Sydney are *London* sailors Peter Seaborn (centre) and John Reekie (right) with a Royal Marine pal. *P. Seaborn Collection.*

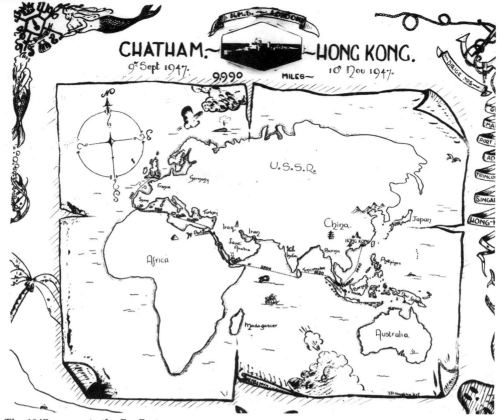

U.S.S.R.

China.

Japan

HONG KONG

Philippines

Germany

France

Spain

Turkey

Iraq Iran

Saudi Arabia

India

Burma

Africa

Madagascar

Australia

The 1947 voyage to the Far East. *Alan J. Howlett Collection.*

packed with passengers on these trips that our hammocks were slung impossibly close together. You could hardly get in or out of them. During the Singapore trooping trip I was accorded the dubious 'honour' of being in charge of Japanese POW working parties engaged in cleaning ship! As a youngster, faced with these 'sons of Samurai' I was more than a little bit nervous.

After decommissioning, the *London* was flagship of the Reserve Fleet for a short period, among obsolete battleships and worn out cruisers that were never likely to see service again.

HMS *London* calls in at Malta on her way East. *Alan J. Howlett Collection.*

While, for many warships, the next step might have been the scrapyard, HMS *London* was, before the end of 1946, redeemed from Reserve and sent for refit at Chatham. In one of the most severe winters in history, sailors living aboard *London* had to endure extreme conditions. Some 'hostilities only' ratings, including Peter Seaborn, were seriously ill, even though they might not know it until after they had left the Service:

I was demobbed during the refit and it later transpired that I had succumbed to Tuberculosis and its accompanying problems. It takes little imagination to understand why there were many others who no doubt suffered the same, bearing in mind our far from ideal living conditions.

Commander R. F. Leonard joined HMS *London* during the winter of 1946-47 and also found conditions hard to bear:

...I have some hideous memories of bitter cold, thick snows and frozen shore-side heads, and a ship apparently disintegrating into a squalid horror of red-lead, rust and wet dirt.[1]

However, the dockyard authorities and neighbouring naval vessels did their best to provide heat and light whenever they could.

The last two years of cruiser *London*'s active service life were to be spent in the Far East, for, in August 1947, she was assigned to replace HMS *Belfast* in Hong Kong as the flagship of the Commander-in-Chief, British Pacific Fleet, Admiral Sir Denis Boyd.

While her extra 20-mm guns had been removed at the end of the Second World War, *London* emerged from her 1946-47 refit much the same in terms of armament as she had been in the war. She did get a laundry, and a theatre to entertain the crew was created in one of the hangars that had not been used for aircraft since 1943. The cruiser also received some small measure of 'air conditioning' and offices were created for the C.-in-C.'s staff.

Originally due to sail for the Far East in June, *London* was damaged during a collision with a dockyard tug and ended up departing in September, after repairs.

Lieutenant Commander Tommy Catlow joined the cruiser towards the end of the refit. A submariner who had spent time in Colditz prisoner of war camp during the war, he was impressed with the *London*, which he found to be '...a successful and happy ship. She was back to pre-war standards of smartness in every respect.'[2]

Prior to the ship leaving, Lieutenant Commander Catlow attended a damage control course, where instructors passed on some valuable lessons from the recent conflict:

It was well presented, and in the not too distant future I was going to be really grateful that I'd done it; I was going to need nearly everything they had taught me.[3]

Fuelling, ammunitioning and machinery trials went without mishap and *London* then set sail for China. The trip was vividly recorded by Lieutenant Commander Leonard who recalled:

...well behaved libertymen and very ill-fitting white uniforms at Gibraltar... difficulties with the heavy stern wires at Malta, where our two or three weeks of working up was marred by a tragedy when an ordinary seaman was lost from a seaboat... and the arrival at Port Said very early in a clear beautiful morning. Of our passage through the Canal, and on to Aden, I remember the discomfort of the anti-fly precautions we took because of a cholera scare...the brilliant naked landscape of the Gulf of Suez... the migrating swallows and the great birds that preyed on them... the requests for 'hands to bathe'... the grimly beautiful movements of the sharks in the clear water when we stopped to fish. From Aden to Singapore I remember only the bright sunset... the peaks of Ceylon against a morning sky... the greenness of Trincomalee... At Singapore we were on our own station and it only remained to paint ship, in the face of the inevitable evening rain, and make an uneventful passage to Hong Kong. And at Hong Kong, I

Aboard HMS *London*, Hong Kong 1947. *Victor Parker Collection.*

An Army officer on *London* spots her fall of shot, with *Sussex* firing in the background. *Victor Parker Collection.*

remember, we made an immediate good impression – which is just what we had come ten thousand miles to do.

A few weeks after arriving in Hong Kong HMS *London* played host to some soldiers who enjoyed an action-packed couple of days at sea. One of them was young Lieutenant Victor Parker of the 2nd Battalion, 'the Buffs', attached to the Gurkha Brigade:

> *We boarded the ship in the afternoon at Tai Po harbour and, after being shown around the ship a film show was given in the evening. She moved off at 06.30 in the morning and sailed out into the South China Sea, escorted by the cruiser HMS* Sussex *and several destroyers. During the day HMS* London *fired her AA guns and also her 8" guns. The target for her 8" guns was a captured Japanese destroyer which, as far as I can recall, soon sank. Night firing was also carried out and was even more spectacular. We spent the night aboard and returned to Hong Kong harbour the following day.*

The ship's cat, 1948-49.
Alan J. Howlett Collection.

Three Cruises

During her final deployment, the *London* would conduct three cruises. She embarked on the first – a 'Southern Cruise' – in February 1948, after working up and exercises. Carrying the Commander-in-Chief, her ports of call were: Saigon, Singapore, Labuan Island, Jesselton in Borneo and Manila. She was initially in company with the cruiser HMS *Sussex*, plus several destroyers and frigates but, after some night exercises, *London* sailed on with the destroyers *Constance*, *Consort* and *Concord*, and the frigate *Alert*, while the other warships returned to Hong Kong.

As the cruiser made her way up river to Saigon, Commander Leonard noted that the pilot seemed in a bit of a hurry:

> *...or perhaps it was just his French temperament. Anyway he increased speed to 22 knots and we hurtled up the river as if possessed of the devil. We cut all the corners; we paid no attention to the native river craft; and half the riverside dwellings must have been swamped by our wash. Not until we came in sight of Saigon did the pilot slacken speed. Our Captain and Navigating Officer admitted afterwards that they were a little shaken by this display of Gaelic verve.*

The potential threat, posed by communist guerrillas fighting to liberate Indo-China

During the late 1940s the Royal Navy experimented with an anchorage at Ominato in Japan. Here, *London*'s **sailors look on as shipmates win Ominato's Fleet Regatta.** *Derek Overton Collection.*

The *London* in refit at Singapore dockyard. *C.W. Ellis Collection.*

HMS *London* in floating dry dock at Singapore, September 1948. *Alan J. Howlett Collection.*

from the French, lay behind the pilot's haste. For Lieutenant Commander Catlow, the Saigon visit was an uneasy experience:

> *Although the French in Saigon were trying to live normal lives, there was an atmosphere of grave unease. Existence was punctuated by occasional explosions in the city as terrorists blew up a building here and there.*

To try and relax he went for a round of golf on a course next to Saigon's international airport.

> *I felt uneasy playing there because I'd heard that quite often the Vietnamese would attack the airport by day. There were a lot of armed French soldiers patrolling the course, and it was difficult to concentrate on the game and keep your eye on the ball.[4]*

The *London* embarked on her next tour in early May. Two destroyers and HMS *Alert* accompanied her initially on this 'Northern Cruise' but after some exercises she was left with only the frigate. On 8 May *London* dropped anchor at Woosung, in the mouth of the Yangtze, and on the following day began her journey up river. Two days later she reached Nanking where the small British community was particularly keen to entertain the sailors. Commander Leonard was bemused by the chaos and confusion of sprawling, anarchic China:

> *It was difficult to think that this city was the Capital of China. Although an ancient walled city, it seemed to us extremely dirty and untidy, resembling a straggling overgrown village rather than a capital city.*

After five days at Nanking, *London* went back down the Yangtze to Shanghai. The cruiser tied up to one of the buoys in Battleship Row, opposite the spectacular Bund waterfront district with its towering European-style skyline. While it appeared more familiar, Shanghai was also an assault on the senses and Commander Leonard was among those dismayed by the decay since its pre-war heyday:

> *Those who remembered the pre-war Shanghai, when the city was an International Settlement, and lovely gardens extended along the full length of the Bund, were shocked to find mud and dirt in place of the gardens, and people and traffic so congested that it could take ten minutes to cross the road.*

But, there was still a large European community in Shanghai, together with an American presence, so the *London*'s crew did not want for hospitality or familiar diversions. On 27 May the cruiser sailed for Formosa (Taiwan), her next port of call, and then headed back to Hong Kong. After a few weeks occupied by sporting activities ashore and some exercises at sea, *London* embarked on her third cruise, carrying Admiral Boyd to Japan in company with *Sussex* and some destroyers. They arrived at Yokosuka Naval Base on 28 July for a five day visit. After a series of exercises with the US Navy, *London* landed the Admiral at Hong Kong and sailed for Singapore, where she went into a floating dry dock to have her bottom scraped. Arriving back at Hong Kong towards the end of November, a series of exercises followed.

In mid-December, HMS *Belfast* arrived and this was the cue for the *Sussex* to return to the UK to decommission. With *Belfast*'s return HMS *London* was sent up to Shanghai, arriving just before Christmas. She was on the threshold of her most dangerous journey.

Notes

1. HMS *London*'s 1946-1949 Commission Book, an account written by Commander R.F. Leonard.
2. Captain T.N. Catlow, *A Sailor's Survival*.
3. ibid.
4. ibid.

Chapter Sixteen

STORMING UP THE YANGTZE

Between a Rock...

The need to confront Japanese invaders during the Second World War had forced the Communists, under Mao Tse-tung, to form an uneasy alliance with the Nationalists. However, with the defeat of Japan, this marriage of traditional enemies quickly fell apart, leading to a vicious civil war in which the Nationalist government forces were unable to hold back the Communist tide flowing down from Manchuria. By 1949 the situation had become so unstable that Royal Navy warships were stationed at Nanking, standing by to evacuate British nationals. Cruisers pulled the same duty at Shanghai, with *London* dropping anchor there every other month.

By April 1949 the Nationalists and the Communist Peoples' Liberation Army were facing each other across the Yangtze, with the latter poised to take Shanghai and Nanking. The Communists, who held the north bank, were known to be planning a crossing of the river in force, to seize the south bank from the Nationalists. To avoid having their movements misconceived, any British warships on the river were trying to get to and from their guard stations before this happened. The destroyer HMS *Consort* was the Nanking guardship, with the frigate HMS *Amethyst* sailing up river to relieve her on 20 April. Even though the *Amethyst* was obviously a British naval vessel she was still fired upon by Communist artillery, sustaining serious damage to her bridge and wheelhouse. Out of control, she ran aground at Rose Island.

As these dramatic events unfolded, HMS *London* was arriving at Shanghai, to take part in St George's Day celebrations, carrying Vice Admiral A.C.G. Madden, Second-in-Command of British naval forces in the Far East. The *London* received a situation report from the *Amethyst*, its details having a 'startling effect' on the cruiser's Commanding Officer, Captain Peter Cazalet. After discussions with Vice Admiral Madden, Captain Cazalet informed the ship's company of the possible implications and *London* was cleared for action. Three pilots capable of taking her up the river were embarked, one of them British, at the insistence of Captain Cazalet:

The *London* in the Far East, 1949. *Terry Potton Collection.*

HMS *London*'s ships company in the late 1940s. *Terry Potton Collection.*

HMS *Amethyst*. *Strathdee Collection.*

HMS *London*'s Battle Ensign flying on the Yangtze. *Terry Potton Collection.*

Feeling that in the event of any shooting the Chinese pilots might not be entirely reliable I asked Mr Sudbury to come on board. He agreed readily and cheerfully to come up the river in the ship; his presence on the following day was invaluable.[1]

The *London* anchored at Kiang Yin, about thirty miles down river from the *Amethyst*, at around 8.30 p.m.

Events had moved on during the afternoon, Consort had steamed at full speed from Nanking to Amethyst's assistance. She too had been heavily fired on, and suffered damage and casualties. She was forced to abandon her attempts to tow Amethyst off and came down to Kiang Yin. She, and Black Swan, who had come down from Shanghai, secured alongside us. Both ships were fuelled and our Medical Staff spent a busy night attending to Consort's wounded.[2]

One of those looking after *Consort*'s casualties was nineteen year-old Sick Berth Attendant, Rod Saul, who contemplated what tomorrow might bring with some trepidation.

As *London*'s damage control officer, Lieutenant Commander Catlow went aboard Consort to see what assistance could be provided:

She looked to be in a hell of a mess: shell holes everywhere, the wheelhouse shattered...Smoke was pouring out of the ship... In many places the ship had been penetrated by anti-tank shells - devastation seemed to me widespread.[3]

At 6.15 a.m. the following morning, *London* weighed anchor and, in company with *Black Swan*, sailed a further ten miles up the river before pausing while attempts were made to clarify British intentions.

In the meantime, *Amethyst* had managed to float off the mud bank but had dropped anchor to avoid provoking the Communist gunners. With efforts to gain some sort of ceasefire guarantee having failed, Admiral Madden and Captain Cazalet agreed that an attempt should be made to prise the frigate out of her predicament.

The *London* was about to embark on a mission against the odds; as illogical in pure operational terms as any of the Arctic convoys. Steaming a cruiser 'past determined and well trained shore batteries in confined waters without prolonged and heavy preliminary bombardment is not a sound operation of war'.[4] But help was urgently needed and the Royal Navy was determined to provide it.

It was unthinkable that London *would abandon* Amethyst *to the Communists and just slink away.[5]*

Hopes for success resided in the fact that the Communists might have realized their error in firing on a Royal Navy warship the previous day. To make it absolutely clear that HMS *London* came only to offer peaceful assistance, a white flag flew from her foremast and the ship was draped in a number of large Union Jacks. At 10.26 a.m. the cruiser weighed anchor

The *London* makes her nationality very clear by displaying as many Union flags as possible. *John Parker/Don Chidlow.*

and, with *Black Swan* following behind, proceeded at twenty-five knots up the Yangtze, which is broad - over a mile wide on some stretches - but also quite shallow in parts. Both warships kept their main armament pointed fore and aft to avoid providing provocation. The Captain was on the bridge, while the Executive Officer, Commander John Hodges, was

One of *London's* 4-inch gun crews shortly before the cruiser went up the Yangtze to try and rescue the *Amethyst*. *John Parker Collection.*

The *London* storms up the Yangtze with guns blazing. *Illustration by Dennis Andrews.*

in the Emergency Conning Position (ECP), situated behind and above the after 4-inch guns. The Admiral and his staff were in the ship's unmanned Y turret, which acted as an auxiliary command position during the action. The medical staff made their dispositions, with Rod Saul sent to the upper deck:

> *The Surgeon Commander, Surgeon Lieutenant, CPO and PO remained in the Sick Bay. One Sick Berth Attendant was to carry out First Aid where needed below decks while one SBA (myself) was to perform first aid duties on the upper open decks. Being in an exposed position soon made me realise that enemy shells do not make exceptions for medical personnel.*

The Communists opened fire at 10.36 a.m., Captain Cazalet calmly noting the likely calibre of their guns:

> *The ship was hit immediately by projectiles which appeared to be of 75mm and 105mm calibre...The firing continued for four minutes in spite of heavy and accurate fire from the whole of* London's *armament.[6]*

Acting Sub Lieutenant Christopher Parker-Jervis, who was with the Executive Officer in the ECP later recalled:

> *Our reactions were very clear – down white flag – train all guns abeam and open fire. We had to wait for them to fire to provide a point of aim, as there was very little to see. As the range was only 2,000 yards or so, we were firing almost horizontally if not at depression, which meant that any errors in elevation, especially for the 8-inch guns, caused very large alterations in range in such a flat and low-lying area.*

Boy Seaman Terry Potton hurried to his action station:

> *I was the most junior in my area so I did quite a bit of running around to the galley to get tea and sandwiches for the others. While I was doing that I heard the noise of battle and got back to my station as soon as possible.*

Marine John Parker was on one of the two starboard twin 4-inch guns:

> *When we closed up for action that morning only the starboard 4-inch and the 8-inch turrets were manned as they knew the trouble would come from north bank that was in Communists hands. I was a loader – you rolled the shell onto the fuse machine where the chap would be in touch with the Gunnery Officer who would tell him how many yards to set the fuse. He would*

use the handle on the machine to do it. My job was to take it off him and put it in the gun. By and large the guns were fired one then the other.

A senior rate on the Admiral's staff helped pass ammunition:

It is difficult to describe the scene, all I can remember was the feverish haste with which we grabbed shells from the hoist (damned heavy too!... and it's not exactly wise to drop them) fused them, rushed them to the gun or ready-use locker... back again... backwards and forwards all the time – smoke and cordite fumes in your nostrils, the incessant Bang! Bang! Bang!... the staccato crack of the close range weapons... the whining of shells and humming shrapnel... the quivering of the whole ship as our guns fired, or we were hit, which was often. Then it was over for a while and we rushed around clearing up decks for the next lot...[7]

The Communists were firing 4-inch and 40-mm weapons, the latter employing high velocity anti-tank rounds capable of piercing all but *London*'s armoured belt. While dug in, the larger calibre batteries were easily spotted, but the 40-mm weapons were well concealed and very difficult to hit. They inflicted the most pain on *London*.

Each time the pattern was the same – a burst of fire from the Bank, quickly followed by our return fire. It was not a pleasant action to be in: the range was never more than 1,500 yards and hits were frequent and inevitable.[8]

But *London* weathered this storm and pressed on. At 11.04 a.m. the cruiser was fired on again and a shell hit the compartment next door to Lieutenant Commander Catlow's Damage Control HQ.

We were unhurt but shaken...the report came from the forward magazine that there was a fire. I knew that if I had to flood a magazine it would increase the ship's draught so much we would probably be aground in the river and if I didn't the ship would blow up. It was not a decision to delay, but luckily this report was followed by a repeat call that the fire was adjacent to the magazine and had been dealt with.[9]

Able Seaman Gunner Charlie Breach was in the compartment penetrated by the shell:

HMS *London*'s 4-inch guns fire on communist shore batteries. *C.W. Breach Collection.*

I was in the lower shell room, loading shells on the hoist to go up to B turret. This Chinese shell came inboard into our compartment and everything went black. There were eight of us in there and only three came out alive. A damage control team opened the door and I crawled out with the other two. But then me and this other bloke were told it was on fire down there and we had to go back down and extinguish it! So we went back down, but luckily it had burned itself out. I had come through the war without a scratch, so it would have been ironic if I had bought it then.

On the bridge Captain Cazalet was increasingly concerned about the steadily mounting casualties:

It was evident that damage and casualties were becoming heavy and I started to consider the advisability of withdrawal. I had in mind the fact that, even if it were possible for London *to reach* Amethyst's *position, the chances of a successful return with* Amethyst *at slow speed were almost negligible.*[10]

Royal Marine Bandmaster Frederick Harwood was assisting the cruiser's medical teams. He was asked to go down to the lower steering position where a damage control team was fighting a fire and there were casualties. He entered a hellish place:

What a mess! Men were fighting the fire by secondary lighting, hardly able to see very much. There was a young lad on the wheel conning the ship, listening to orders on the voice pipe, surrounded by men putting out the fire. He was very brave. What I did not realize at the time was the main control of the rudder had gone and that all of us were in his hands. We were in a river, which was getting narrower and all of us were relying on him, one young lad. In the dim light I saw someone on the deck. He touched my leg. I spoke to him, and he told me his leg was gone. It was now getting unbearable down there so I went to the hatch, and they passed me a stretcher. I felt his leg but could not see; I got two handkerchiefs, tied them together, and put them around his leg. I got a piece of wood, twisted it round, and put the end of the wood under his belt. I got him up the hatch, took him to the PO's Mess and laid him on the table. I had a word with him and he was still conscious. I told him to wiggle his toes; he did, or thought he did, and then passed out. The Mess was full of wounded so I had a quick word with those that wanted to talk and lit their cigarettes. Then I went back to the Sick Bay.

Aboard the *Amethyst* they heard the gunfire and got ready to follow *London* back down the river once the cruiser broke through.

...this got louder and louder and we thought any time now and we should get under way... But, then slowly the gun fire seemed to get further and further away and we eventually realised that they had been so badly damaged trying to get to us that they decided it wasn't possible for them to come up and rescue us.[11]

At 11.06 a.m., when *London* was about nineteen miles from *Amethyst*, a shell had burst over her bridge, just as the decision to turn around had been taken by the Captain:

Damage to instruments and communications on the bridge was obviously severe and I was in doubt as to whether I was still in control of the ship... I considered that this passage at high speed under heavy fire could not be controlled from the Emergency Conning Position abaft the funnels. I was in no doubt that the time for withdrawal had arrived and ordered the wheel to hard astarboard and the starboard engine to full astern.[12]

Lieutenant Neil Stewart, who was the engineer of the watch in the forward engine room during the action, took the order:

We could hear the Communist shells clanging on the ship's four inch armour and also feel and hear our secondary and main guns replying. Because telegraphs were not working

I got the order by telephone. We must have been going at least 20 knots when they said full astern. When the engines go full astern they put the helm over. If we had still been going ahead we would have gone right up on the bank. We could have been in real trouble, but the wash from the screws and the fact that we were going full astern when we turned around is what saved us.

As third in command, Lieutenant Commander Catlow was called to the bridge where he was confronted with 'not a pleasant sight'.

It killed the Chinese pilot and knocked out the Captain and Rowan Grice-Hutchinson, our navigator, who later died. Mr Sudbury, the civilian pilot, was unconscious; Sub Lieutenant Haggie, Officer-Of-the-Watch, was conscious and still upright but not 'with us'. The ship had the wheel hard over and we were going at about 20-25 knots for the north bank... .[13]

At the same time as Lieutenant Commander Catlow arrived on the bridge so did the Fleet Navigating Officer, Commander Hare, a member of the Admiral's staff, and the Executive Officer, who had come up from the ECP.

About that moment we hit the north bank of the river and swung back into the main stream.[14]

With officers more senior to him now on the bridge, Lieutenant Commander Catlow returned to the Damage Control HQ. Marine Parker and the other 4-inch guns crews now had to make a mad dash for the other side of the ship:

When the ship turned around the other crews weren't closed up so we had to get quickly through from the starboard 4-inch guns to man the port ones. It was pretty tricky underfoot as, with the ship turning, all the shell cases were rolling across the deck.

Below decks Bandmaster Harwood sensed the warship was making a less than graceful about turn.

We felt the ship's bow rise and then we hit the sandbank. Then full astern, and we were on our way down the river... .

The Captain handed control over to Commander Hodges while he had his wounds attended to. Mr Sudbury, having regained consciousness, had taken over as pilot, the surviving Chinese pilot having retreated to Y turret.

During the passage down river the ship was fired on by five batteries, three of which had not fired when previously passed. Damage and casualties mounted steadily and were considerably heavier than those sustained on the passage up river.[15]

Chaos reigned in the 4-inch gun positions where the senior rating from the Admiral's staff just had time for a cigarette before all hell broke loose.

Back down the river, all guns still in action... hole blown through our messdeck, the funnel, the hangar, the cinema; twisted metal and powder stains everywhere... Fire on the 4" gun deck, spreading to the shell lockers, got it out in time and ditched the red hot live shells in the drink before they went off... No2 on the gun wounded, took over until relief arrived... A lull for a minute – stagger around with shells again – my God we looked like a crowd of filthy tramps... ripped clothes, blackened faces, still grinning though...Very hot in Action Working Dress and anti-flash... someone brought some apples and oranges up... managed to get half an orange before it became advisable to seek cover... eat a bit more orange while half under a lorry on the gun deck... terrific explosion... dropped orange, legs felt numb... someone crying out, screams filled the air... felt slightly hurt at dropping my orange. Moved my legs OK... no injuries I think. Willie lay by me, all blood... I was covered in blood... not mine... Bob lay moaning 'my leg, my leg!' The Chief was by me on his stomach, he looked alright... but he died before we got him to

At the dockyard in Shanghai, one of *London*'s Royal Marines shows the cruiser's scars to a visitor. *Terry Potton Collection.*

the Sick Bay... *Two of us carried a bloke down, the firing was almost stopped... there was nobody to fire our gun anyway – the Reds wrapped up too. 'Thank God!'*[16]

Marine Parker had somehow come through it all without a scratch:

I had been under fire during the Sicily landings in World War Two, but this was something else altogether. Amazingly I didn't suffer any wounds at all. I was very lucky because the hit that did all the damage struck the side of the ship and sprayed shrapnel everywhere. I remember at one stage we ran out of High Explosive ammunition so we loaded starburst shells and fired them – the Chinese must have wondered what the heck was going on as this harmless fireworks display erupted over their heads.

Damage Control became almost impossible and, as Lieutenant Commander Catlow supervised his teams in filling up holes on the waterline with hammocks, he became a little punch drunk:

We were in a real mess. There had been many fires, including one in the paint locker; paint burns well, but it was floodable, so with some misgivings, because I heard that our Chinese mess boys might be down there, I flooded it. I'm glad to say that the Chinese had been prescient and gone somewhere else. During my walking round the ship I came on the padre comforting a dying man, and I found

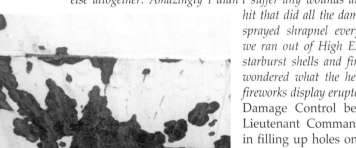

Hammocks stuffed in a shell hole in *London*'s side to prevent water coming into the ship. *Derek Overton Collection.*

myself saying to him that he ought to be congratulating, not comforting him, because he was going to a better place. At that time I must have been very slightly shocked, because I went into the sick bay and, finding my colleague and friend Lieutenant Bob Ingham on the operating table having shrapnel taken out of his backside, just stood and watched without any feeling of revulsion until I suddenly became myself again and went about my business.[17]

The senior rating from the Admiral's staff was also in the sick bay, doing his best to comfort wounded shipmates:

A communist shell hit *London*'s bows, but failed to penetrate. *Derek Overton Collection.*

Bob was bearing up with a smashed leg... I talked to him and promised to send a telegram to his wife... A young Ordinary Seaman named Warwick held my hand, told me it was his first ship, said he'd never be able to play football again... I told him he would be alright in a few weeks – I was wrong, he died. Ron and I, the only two (almost) that came off the 4" guns under our own steam, did a proper Florence Nightingale act – rushing around, scrounging fags, tea, water, dishing this out to the lads – doing what we could for those in pain... we had no time to think, luckily, and when we finished sat down on the deck and went to sleep.[18]

The Communists had stopped firing at 1.40 p.m, the *London* having been under fire for forty-eight minutes spread over a period of three hours. She had suffered thirteen killed, fourteen seriously wounded and forty lightly wounded, with the exposed 4-inch gun crews and their supply parties taking more than a third of the casualties. But, despite their losses, the 4-inch guns had kept firing, casualties being almost instantly replaced.

Ultimately the Yangtze Incident, as the episode became known around the world, claimed fifteen lives in *London* and forty-six people were killed altogether in *Amethyst, Consort* and the cruiser. There were no fatalities in *Black Swan*.

In total the *London* fired 155 8-inch shells, 449 4-inch and 2,625 rounds of close range ammunition.[19]

The Communist Noose Tightens

Back at Shanghai, HMS *London* went alongside at Holt's Wharf across from the dockyard. Lieutenant Commander Catlow, who had become temporary Executive Officer while Commander Hodges was temporary Commanding Officer, supervised the clear up:

There were still small fires smouldering on board and these were dealt with. The aftermath of battle is appallingly squalid. Dockyard personnel started to work at once, making us watertight by welding plates on the ship's side to cover shell holes... The damage on close inspection made me wonder whether it would be worth mending – after all, she'd had a modernising refit but was now over 20 years old. Although the shell holes in the ship's side were bad enough, we had suffered internally devastating damage from solid anti-tank bullets...I found one of those bullets

A funeral in Shanghai for *London*'s dead. *Don Chidlow Collection.*

sticking through the ship's port side. It had gone through the starboard side of the ship, right through the turret barbet of B Turret, and the machinery...

After surveying the destruction, Lieutenant Commander Catlow felt rather weary and upset:

I sat in my cabin and reflected on the horror of the day. It was at this moment that the enormity of our losses really hit me and I felt desperately sad at such a waste of young men. The whole thing had been a dreadful misunderstanding. There had never been any intention to interfere with the Communists' crossing of the Yangtze – that was their business.[20]

Conducting his own inspection of the battered ship, Boy Seaman Terry Potton concluded he had been very lucky:

What made me a shudder was seeing the side of the ship with hammocks sticking out of shell holes at a level similar to my action station. They had proposed sending landing parties ashore to try and secure the area along the Yangtze. I am glad that idea was abandoned as I think it would have been suicidal.

Another of *London*'s young sailors was amazed that a matchstick model of the cruiser, made in Hong Kong and presented for display in the communications Petty Officers' mess, had somehow come through the action unscathed.[21]

Visits by some of *London*'s sailors to a Shanghai hospital brought home the dreadful human cost. One of those hovering between life and death was navigating officer Lieutenant Commander Grice-Hutchinson, who was to die on 6 May, after being transferred to a hospital in Hong Kong. Those that died on the ship were buried at Hung Jau Cemetery and a packed memorial service was held at Shanghai Cathedral the following day.

The cruiser's repairs took a week and, as Shanghai emptied around them, her crew were anxious to sail for Hong Kong. Marine Parker was one of those who feared the worst if the *London* didn't make a break for it soon:

The buzz did go around that if we were still there when the Communists took the city they would impound the London.

But, still the Royal Navy delayed *London*'s departure for a few more days, in order to avoid alarming the rapidly diminishing local British population.

However, we couldn't afford to leave it too late because about 15 miles down river, where the Whangpo meets the Yangtze, is a comparatively narrow exit with fortifications on both banks of the river. If these were manned with evil intent by the Communists, the ship would be sunk without any doubt... .[22]

In the meantime, a number of *London's* sailors began to feel the stress and were reporting sick with psychosomatic illnesses. Their experiences had left them feeling mentally fragile and their anxiety leapt when it looked as if the *London* might have to fight her way out to the open sea. Communist troops were entering Shanghai and had possibly occupied the forts at Woosung. As she started her journey down river the cruiser went to action stations with all guns ready to fire. But, as Britain was not at war with the Communists, the *London* could only fire in self-defence.

The guns of the forts were pointing towards the centre of the river from both sides but we could see no signs of life. As we passed between the forts it was a very eerie feeling that they might just blow us out of the water; but nothing happened ... .[23]

The decision to sail up the river and try to rescue *Amethyst* was a matter of conscience and after PQ17 the Royal Navy, and in particular HMS *London*, knew that trying in spite of the odds was better than not trying at all. The decision was left to the commanders on the spot who were not strait-jacketed by interference from the Admiralty.

Homeward Bound

After a welcome fit for heroes at Hong Kong, the *London's* sailors began to look forward to their imminent return to the UK.

Our happy anticipation during that time was clouded by the thought of the Amethyst, *still in her perilous and comfortless position in the grip of the Communists.[24]*

The *London* departed on 14 June, sent on her way by a tumultuous send-off from the ships of the Far East Fleet, their crews cheering her out to sea.

HMS *London* in the Mediterranean on her way home to the UK. *Strathdee Collection.*

After spending a few weeks based in Singapore, *London* finally set course for home, flying her paying off pennant from her foremast. She went with a glad heart.

> *When the news of* Amethyst's *brilliant and gallant escape came through on the 31st July, it brought a greater thrill to none more than us. Now we could go home completely happy – the Yangtze Incident was finally and satisfactorily closed.*[25]

Soon *London* was steaming past Sabang where nearly four years earlier she had taken the Japanese surrender. After topping up with oil and water at Aden, the *London* headed up the Red Sea and entered the Suez Canal. Marine Parker was amazed by the reception the British Army gave his ship:

> *By the time we came to pass through the Suez Canal we were quite famous due to our exploits on the Yangtze. There were troops the full length of the canal. Some of them were getting in trucks and then racing ahead of us so they could cheer us through again. It was like a relay race.*

After a rough crossing to Malta, and a brief stop to send mail ashore, the *London* went non-stop to Gibraltar.

On 5 September the Isle of Wight was sighted and by dawn the next day the Great Nore Tower was on the horizon. After a couple of days at Sheerness, where the ship was de-ammunitioned and de-oiled, she finally headed for Chatham.

> *We liked the idea of being officially welcomed by the Lord Mayor of London, but we enjoyed even more meeting our families and friends... And then, of course, we had some foreign service leave to look forward to – in our own homes. The trip home took just over a month. The sun shone practically all the way and we arrived back nicely sun-burnt with tans ranging from pale puce to deep mahogany. The tan will soon wear off. But not the memory of the 'Lovely London'.*[26]

On 16 November 1949 *A Service of Thanksgiving To Almighty God for preservation from danger in the River Yangtse* (sic) was held at the church of St Martin-in-the-Fields, London.

The following day, those who were to receive awards for their bravery during the Yangtze Incident went to Buckingham Palace, where King George VI presented them with their medals. As he pinned a Distinguished Service Medal to Bandmaster Harwood's chest the King said: 'Thank you for looking after my men.'[27] The Bandmaster's citation said that he had gone sixty hours without sleep to attend to the wounded during and after the action. The Senior Gunnery Officer, Reginald Smith, received a Distinguished Service Cross, Captain Cazalet a Bar to the Distinguished Service Order he had won on the Arctic convoys and Able Seaman Alan Dudley the Distinguished Service Medal.

A memorial plaque commemorating *London*'s dead. *Terry Potton Collection.*

Surgeon Commander W.B. (Bill) Taylor, Chief Petty Officer Stoker Mechanic Henry Fletcher, Corporal William Hart, Royal Marines and Chief Petty Officer Thomas Learmouth were all Mentioned in Dispatches. Chief Writer Patrick Joseph Stowers received the award post-humously. Another thirty-one members of *London*'s ship's company were mentioned in a Special Order of the Day from the Commander-in-Chief of the Far East Station, for their 'courage and devotion to duty'.

In the meantime, the *London* had been reduced to Reserve and in early 1950 was sold to T. W. Ward for scrap, arriving at Barrow-in-Furness in the north-west of England for breaking up on 22 January.

Notes

1. Papers of Vice Admiral Sir Peter Cazalet, Imperial War Museum.
2. HMS *London*'s, 1946-1949 Commission Book, an account written by Commander R.F. Leonard.
3. Captain T.N. Catlow, *A Sailor's Survival*.
4. HMS *London*'s 1946-1949 Commission Book, an account written by Commander R.F. Leonard.
5. Captain T.N. Catlow, op.cit.
6. Papers of Vice Admiral Sir Peter Cazalet, Imperial war Museum.
7. An anonymous account of the action, in the papers of Bandmaster Frederick Harwood.
8. HMS *London*'s 1946-1949 Commission Book, an account written by Commander R.F. Leonard.
9. Captain T.N. Catlow, op.cit.
10. Papers of Vice Admiral Sir Peter Cazalet, Imperial War Museum.
11. Keir Hett, Sound Archives, Imperial War Museum.
12. Papers of Vice Admiral Sir Peter Cazalet, Imperial War Museum.
13. Captain T.N. Catlow, op.cit.
14. ibid.
15. Papers of Vice Admiral Sir Peter Cazalet, Imperial War Museum.
16. Papers of Bandmaster Frederick Harwood.
17. Captain T.N. Catlow, op.cit.
18. Papers of Bandmaster Frederick Harwood.
19. HMS *London*'s 1946-1949 Commission Book, an account written by Commander R.F. Leonard.
20. Captain T.N. Catlow, op.cit.
21. C. Clifford in a letter to the author.
22. Captain T.N. Catlow, op.cit.
23. ibid.
24. HMS *London*'s, 1946-1949 Commission Book, an account written by Commander R.F. Leonard.
25. ibid.
26. ibid.
27. Quote recalled by Frederick Harwood in interview with the author.

Chapter Seventeen

THE LONG RETREAT

A Product of the Missile Age

Without the money to build new escort vessels, except half a dozen Daring Class destroyers, the British fleet was forced to convert a number of Second World War warships to Anti-Submarine Warfare (ASW) and Anti-Air Warfare (AAW) frigates in the early 1950s. This was in response to Soviet submarine and maritime strike capabilities which, although crude, were expanding rapidly.

Despite its much diminished state, the Royal Navy was, by the late 1950s, able to find enough funds to keep itself at the cutting edge of warfare technology, with a class of guided-missile destroyers that were the first of their kind in the British fleet. The County Class programme had been initiated in the late 1940s to provide escorts for the remaining fixed-wing aircraft carriers. It aimed to create a warship with leaner manning and more comfort, armed with the latest weapons and sensors, which was capable of surviving and fighting through nuclear fallout. The new Counties had 'a lightweight superstructure designed to best attack and defend... by manoeuvrability and speed rather than armoured protection...'[1] The tenth HMS *London* was the third of the class and was laid down at Wallsend shipyard on the Tyne, towards the end of February 1960. Built by Swan, Hunter & W. Richardson, she was launched on 7 December 1961, going alongside the fitting out wharf where her 'immaculate coat of paint... soon acquired a heavy overlay of Tyneside muck'.[2] Sea trials were carried out in April and May 1963.

> *Wearing the Red Ensign,* London *slipped down the Tyne and was put through her paces as a ship. Very little of the equipment, that make* (sic) *her a warship, was in working order at this stage, and it was mainly the hull and machinery that were on trial.*[3]

The new *London* cost around £13.9 million to construct[4] and was commissioned into service on 14 November 1963, 'the newest and most up-to-date ship of war afloat'. Her displacement was 6,800 tons (full load) and she was 520 feet long, with a beam of fifty-four feet. Her crew was composed of up to 505 officers and men.

The guided-missile destroyer HMS *London*. *Illustration by J.S. Morris.*

Fred Fleming was a sixteen-year-old Ordinary Seaman when he joined HMS *London* as she came out of build:

> *To me she seemed like a luxury liner... You had bunks, and you even had televisions in your mess decks, which was pretty amazing in those days. On other ships you would have been in hammocks and had no TV, just a radio in the mess. My job was in the Operations Room, which had its own innovations, including built-in ashtrays on every radar console.*

The tenth *London* was designed with enough speed (32.5 knots) and endurance to keep pace with carriers on overseas deployments. Along with the Tribal Class frigates constructed during the same period, the Counties had Combined Steam and Gas (COSAG) turbine propulsion. The steam turbines handled normal cruising while the gas provided additional power for high-speed bursts. They could be flashed up very quickly, allowing the *London* to leave harbour at short notice if necessary.

Her flagship capability – she had enough accommodation and also the command systems for an Admiral and his staff – combined with guided missiles, meant the new *London* was informally known as a Destroyer Leader-Guided (or DLG). The Seaslug Mk1 surface-to-air missile (SAM) system dominated the design of the *London* and her sisters, necessitating a broad passageway through the ship and a large launcher on her stern.

The *London*'s bows make an impressive sight. *Ian Inskip Collection.*

> *The missiles are brought up from the forward magazine on a hoist and then travel through the tunnel, undergoing checkouts, fitting of wings, etc., before being run out on to the launcher. This was a somewhat over-engineered, massive lattice-work affair weighing as much as a gun turret.*[5]

The DLG's magazine was 285 feet long and the Seaslug hydraulic handling machinery was 'the largest...ever to be seen at sea in a British warship'.[6] Seaslug took care of the long range, high altitude air threat (with a limited capability against surface targets), while the much smaller Seacat SAM was for close range. To handle hostile submarines, *London* carried a Wessex helicopter, equipped with dipping sonar and armed with homing torpedoes. The County Class destroyers also packed a more traditional punch, with two turrets mounting twin radar-controlled 4.5-inch guns forward.

The man in charge of those guns, as well as the Seacat system and the ship's magazines, was Lieutenant Douglas Clark, who was *London*'s 2nd Gunnery Officer between 1963 and 1965:

> *When she commissioned she was one of the most advanced ships afloat. From weapons to radars, it was all new equipment. The type of 4.5-inch turrets she was fitted with had previously suffered problems in the Daring Class destroyers but they were well proven by the time the new* London *got them. I had a sort of family connection with the* London *as my wife's father had served in the three-funnel cruiser of the pre-war years. I was also familiar with the previous* London *because between 1937 and 1941 I served as a Boy Seaman in her sister ship, HMS* Shropshire, *in the 1st Cruiser Squadron, of which the old* London *was flagship in the Mediterranean Fleet.*

Give That Man a Coconut

One of the principal guests at the destroyer's commissioning was Admiral Sir Peter Cazalet, who had taken the previous *London* up the Yangtze, while the tenth HMS *London* herself was commanded by the legendary Captain Josef Bartosik. Bartosik was one of two Polish-born captains serving in the Royal Navy during the early 1960s, both men having preferred to stay in the British fleet, after distinguished service during the Second World War, rather than go home to Communism.[7]

The *London* left the Tyne two days after commissioning for acceptance trials, which went fairly smoothly. However, within a few hours of being accepted, there was a major steam leak in both of the warship's boilers and she lost electrical power, spending the weekend at anchor on the Firth of Forth while contractors rectified the defect. Going around the top of Scotland the following week '...the waves began plucking the brand new paint from the ship's side...'[8] and she arrived at Portsmouth on 25 November looking less pristine than Captain Bartosik might have preferred.

The final sea trial took place in the Moray Firth in April 1964 and then *London* '... set sail for Fishguard to fire our missiles on the Aberporth range. Each individual system had made the grade, but would the whole lot work together when the button was pressed?'[9]

The Seacat and Seaslug performed superbly and the ship scored 100 per cent against target drones; a feat she also achieved in front of the Duke of Edinburgh that June when he visited the ship during pre-deployment work-up at Portland.

> *Unerringly a rocket sped to its target many thousands of feet above the destroyer and scored a direct hit on a pilotless aircraft.*[10]

What happened next became legend. On seeing the target explode the Duke of Edinburgh exclaimed, in imitation of someone reacting to a bullseye at a fairground attraction:

> *Give that man a coconut!*

He was true to his word, as Brian Parker, one of *London*'s junior communications ratings at the time, recalled:

Seaslug is launched. *Ian Inskip Collection.*

The *London* **arrives at Houston, Texas, 1964.** *Strathdee Collection.*

The missile firings were a real bonus for the Duke. About a week later a glass case containing a coconut arrived by courier direct from Buckingham Palace addressed to the Captain. It was presented to the Seacat team at a ceremony onboard.

At the beginning of July the destroyer paid a visit to London, the first warship named *London* to do so since the cruiser. She deployed in mid-September, heading for the Far East, via the Americas and South Africa.

Second Gunnery Officer Douglas Clark had last been in the Far East at the end of the Second World War aboard the cruiser HMS *Newfoundland*.

Conditions aboard the Newfoundland *were terrible – no air-conditioning, and with the ship packed with a full war complement. Everybody suffered dreadfully. But, back in the same part of the world nearly 20 years later, the* London *was an absolute palace compared to* Newfoundland, *with good accommodation and, most importantly, air-conditioning.*

The Confrontation with Indonesia

Although Britain was gradually withdrawing from her empire, commitments would remain through the 1960s and into the next decade too.

In 1965 confrontation with Indonesia was providing a focus for operations. Malaya had been given independence in 1957, after British forces defeated communist guerrillas, and Singapore got its independence two years later. Malaysia was created in 1963 by uniting Singapore, Malaya, Sarawak and Sabah. Indonesia – made up of former Dutch colonies and under the leadership of the pugnacious Achmed Sukarno – objected most strongly to the formation of Malaysia. Indonesia already possessed southern Borneo and had its eye on acquiring Sabah (previously known as North Borneo). One of the main reasons behind the urge for territorial expansion was Sukarno's desire to take over the oil rich sultanate of Brunei, on the northern coast of Borneo. The Indonesians also wanted to incorporate Singapore into their territories. Singapore had withdrawn from the Malaysian federation in 1965 but its security was still guaranteed by Britain along with that of Malaysia and Brunei.

An insurgency, supported by the Indonesian military, provoked a forthright response. The Royal Marines and Special Boat Service played a leading part in jungle fighting while the Royal Navy patrolled offshore. Aircraft carrier task forces operating out of Singapore were a significant deterrent against more overt Indonesian military intervention. The threat to British warships in Singapore was taken seriously enough to warrant submarine nets and other measures, as Engineering Artificer 1st Class Brian Dunster in HMS *London* found out:

> *An awareness of the potential threats was encouraged and, for example, we turned the propellers every now and then while alongside just to deter frogmen who might try to place limpet mines on the ship.*

Brian Parker remembered the same countermeasures:

> *Our own divers were sometimes used to check the ship out while the crew was cleared on shore. I remember the Australian warship HMAS* Yarra, *berthed on the other side of the dockyard, had a close call, with an Indonesian diver found dead floating nearby. They used to drop scare charges near the ships in harbour and I believe that's what killed him.*

Such incidents made *London*'s upper deck sentries very edgy and one night teenage Marine Engineer Ken Fleming looked over the side of the ship and thought he spotted disaster:

> *It looked like a limpet mine attached to the ship, about a foot above the waterline. So, I blew my whistle to raise the alarm. However, it turned out to be a magnetic securing ring stuck on the side of the ship by one of our own divers. He had been doing some sort of maintenance below and had forgotten to take it off.*

The *London*'s sailors thought their time had come in the wee small hours one morning, when three explosions went off under the hull. With klaxons blaring everyone leapt from their bunks and went to their action stations. It was a false alarm caused by a young midshipman lobbing scare charges into the water on seeing phantom frogmen. On dry land the situation was equally tense, with violent pro-Indonesian street protests requiring *London*'s sailors to be trained in the art of riot control. They donned steel helmets and carried shields and sticks.

The *London* was at Singapore until 8 February before sailing for a visit to Port Swettenham in Malaysia and heading north to Hong Kong, Britain's other main naval base in the Far East. The destroyer left Hong Kong in early March, returning to Singapore via Bangkok and thereafter spent eight weeks in port, with only short visits to Hong Kong and the Philippines.

> *The reason* London *spent most of her time in harbour was because she had a major problem while approaching Singapore Naval Dockyard at slow speed. A severe vibration was felt by all onboard and investigations revealed that the 'A' bracket that held the drive shaft externally to the screws had come loose and the shaft had moved in the stern gland. While the ship was in dockyard hands many of the crew were out- posted to other ships including New Zealand and Australian vessels.*[11]

The *London* did, however, manage a few patrols in tense waters. She provided air defence for Royal Navy carriers as they continued to deter the Indonesians. Chief Radio Supervisor Edward 'Tex' Davis was among those who felt his pulse rate leap during an alert:

> *The ship was in the Lombok Strait when we suddenly went to Action Stations – Seaslug had acquired target against a large darkened vessel thought to be Indonesian, only to be identified just in time as the commando carrier HMS* Bulwark.

The *London* headed back to the UK in July after a final visit to Hong Kong. On the way back the destroyer called in at the British controlled port of Aden to provision, with junior rating Ken Fleming putting on his steel helmet again:

> *I pulled some sentry duty on the jetty, armed with only a whistle and a stick. It was a bit dodgy, as the people that wanted British rule to end had the habit of lobbing grenades around. While the ship was in at Aden some potatoes were delivered in sacks so we emptied them onto an escalator on the jetty, to enable us to see if there were any bombs concealed in with spuds. The potatoes went up onto the ship and then we put them back into bags and carried them down into the hold.*

As she headed north to Suez up the Red Sea, *London* went to the aid of an Italian tanker that had run out of water in soaring temperatures, and ended up salvaging the 23,000 ton vessel, which earned her crew some prize money.

The *London* sailed back into Portsmouth in late August, with her first commission officially ending in November. The confrontation with Indonesia in the Far East would last until August 1966, Sukarno having been deposed by a military coup five months earlier.

The Withdrawal From Aden

After a short refit, HMS *London* recommissioned on 20 May 1966 under a new commanding officer, Captain David Forbes, who had at one time served in the cruiser HMS *London*.

The destroyer deployed to the Mediterranean in October and November, visiting Italy, Greece, Malta and Gibraltar. On her return to Portsmouth she took up the role of flagship of Flag Officer First Flotilla (FOF1). Rear Admiral Michael Pollock switched his flag to her from the cruiser HMS *Tiger*, which was paying off for a major reconstruction.

On 13 March 1967, shortly after returning to Portsmouth from an eight week cruise in the West Indies, HMS *London* was unfortunate enough to suffer a fire in her galley.

> *Fire broke out early yesterday while many of the crew were still in their bunks, but no-one was injured. Dockyard firemen, helped by Portsmouth brigade, wore breathing apparatus to fight the fire, which took nearly an hour to put out.*[12]

The fire damaged *London*'s missile control system but, because it was close to the Seaslug magazine, it could have been far more serious, as Bob Boynton, one of her junior ratings at the time, explained:

> *If the magazine itself had caught fire the explosions would have been enormous. If I remember correctly, HMS* Glamorgan *and HMS* Kent *were berthed forward and aft of us, also fully laden with missiles! The Naval Base and a chunk of Portsmouth could have been wiped off the map of UK!*

In November 1967 *London* was involved in providing cover for the evacuation of the last troops from the British colony of Aden – Operation Magister – amid fierce fighting, as neo-Marxists fought with pro-western sheikhs to gain control. The destroyer had sailed from Portsmouth at the beginning of October, going around the Cape, as the Suez Canal was blocked. *London* provided an escort for the carrier *Eagle*, conducting a series of exercises within sight of the Aden coast, in company with half a dozen other warships, a submarine and Royal Fleet Auxiliary support vessels.

Fred Dale had joined the *London* in October 1965 as her Master-at-Arms, not long after a brief visit to Aden while on attachment to the RAF:

Cruising off the Australian coast, late 1960s. *W. Luke Collection.*

> *Aden was not a nice place – going there was a bit like attending a party in a cage full of lions.*
> *I well remember standing on the bridge of* London *watching the last of the aircraft taking off*
> *and thinking 'good riddance'.*

The *London* was at sea non-stop for forty-eight days during her Aden standby, adding another 7,000 miles to her clock, and with some of her Operations Room staff closed up for action throughout. 'It was the longest period of non-stop operations at sea I had known in 34 years of naval service,' Captain Forbes observed as *London* returned to Portsmouth.[13]

The destroyer concluded her commission that April, with a trip up the Thames to Greenwich before entering a six-months' refit.

'Hands to Bathe' in the Black Sea

During 1969 *London* found herself taking part in the Beira Patrol on her way out to, and return from, the Far East. Royal Navy warships were enforcing a United Nations oil embargo on the renegade state of Rhodesia by patrolling in the Mozambique Channel, just off the port of Beira. The harbours of the Orient were far more enticing than East Africa, as Leading Seaman Gunner Warwick 'Jan' Luke soon discovered:

Possibly the Royal Navy's first five ship replenishment-at-sea (RAS), with HMS *London* **third from left.** *W. Luke Collection.*

The London *was a great ship – excellent in every way to serve in. We did a lot of very enjoyable flag showing, going to Australia, New Zealand, Japan, Hong Kong, Singapore, Malaysia and I remember we did missile firings off Roosevelt Roads, Puerto Rico. The food aboard* London *was quite good as we had Maltese chefs. When you were morning watchmen up scrubbing decks, the smell of lovely fresh bacon wafted up through the ventilators. You were first to it and it tasted fantastic. When we went to Tokyo we played a Japanese civilian side at rugby and they were very good but we beat them. One time we were in the China Fleet Club in Hong Kong and this US Marine on R&R from Vietnam walked in. He asked us why we weren't fighting alongside the Americans. My mate said that the Viet Cong were doing a good enough job without our help. It ended in a big scrap.*

Having left the ship in May 1968, Fred Dale returned to *London* for his second stint as Master-at-Arms in April 1970, serving in her until August 1972. He recalled an amusing encounter with the Soviets in the Black Sea during her 1971 deployment to the Mediterranean:

One day I went to see the Captain to request our usual swim. He said that we would have something of a problem as we were in a position whereby we could not fully turn the ship's engines off, for fear of drifting into Russian territorial waters. He instructed me to announce that only strong swimmers were to enter the water and must keep clear of the area near the screws, as they would be turning slowly. This I duly did and at the appointed hour the strong and more enthusiastic swimmers lined the deck. The plunge was made. The CO, Captain Ron Forrest, and I performed our usual little unofficial race. We had swum around about a hundred yards from the ship's side when we both looked up and saw that the Russian destroyer shadowing us had 'hove to' some fifty or so yards ahead of us. Their listening gear had obviously noted that our ship had its engines running and, with the misinformation that the Soviet regime fed to them, surely believed that all the men in the water were deserting the ship and heading for communist Utopia. As we looked up at the destroyer, which by now was only a few yards from us, we saw what seemed to be their whole ship's company lowering scrambling nets and obviously expecting us to join them. Both the Captain and myself turned and made rather frantic signals to the men behind us to abandon 'hands to bathe'. They too had realized the situation, and were already turning and swimming for the safety of the London. *Our enjoyable swim was annoyingly curtailed, but a possible international incident had been avoided.*

Farewell to Malta

In August 1972, *London* entered a refit that lasted somewhat longer than anticipated. The economic crisis of the time, caused by the high price of oil, forced Portsmouth Dockyard to adopt a three-day working week with subsequent delays. Emerging in November 1975, with the latest version of Seacat and improved accommodation, the *London* embarked on sea trials and work-up.

The following summer she sailed for the USA to take part in the Bicentennial celebrations at New York, where some of her sailors marched through the city alongside officers and ratings from the frigates *Bacchante* and *Lowestoft*.

In 1977 *London* played a prominent part in the Queen's Silver Jubilee Review, held on 28 June at Spithead, the last event of its kind in British naval history.

Chief Petty Officer Mike North recalls that the *London* of this period was without doubt a happy ship in which to serve.

The *London*'s 1st XV rugby team pose for a team photo beneath one of her twin 4-inch gun mountings. Warwick 'Jan' Luke is fourth from the right, front row. *W. Luke Collection.*

> *The* London *had a very good ship's company. We worked hard and we played hard. She was a good family ship. My son, Richard, was christened in her ship's bell in April 1978 while the* London *was alongside in Pompey dockyard. It took place in the Captain's cabin, with the bell turned upside down and placed in a special frame, like a font. The padre christened my son using a tiny drop of rum and the cake was cut with the Captain's sword.*

In 1978-79 *London* visited the USA and spent some time on NATO exercises in the North Atlantic and in the Mediterranean as well as visiting the Baltic, including a visit to Gdynia in Poland, which at the time was a Warsaw Pact country.

But the most memorable episode of that period came in March 1979, when *London* was flagship for the final British military withdrawal from Malta, ending a Royal Navy

A good view of *London*'s Seaslug launcher as she comes alongside at Gdynia, Poland, during her 1978 visit to the Baltic. *Mike North Collection.*

Farewell Malta. As flagship for the final British military withdrawal from the island fortress, HMS *London* was the last Royal Navy warship to leave. *Illustration by Dennis Andrews.*

presence lasting 179 years. Malta's socialist government had been inspired to bring the British presence to an end by Libyan dictator Colonel Gaddafi who was invited to attend the withdrawal ceremony. Chief Petty Officer North watched from *London* as the Colonel's flagship pulled in:

> *This cruise ship came in with a load of Libyans onboard who went ashore and paraded through the streets carrying banners with drawings on them showing the Libyan boot kicking the British out. They didn't get a very good reception from the Maltese who chased them back to their rusty old liner.*

One of *London*'s sailors, Radio Operator David Gilchrist, carried out the symbolic lowering of the British flag over Fort St Angelo, at midnight on 31 March. To the dismay of the majority of Maltese attending the ceremony, Gaddafi, together with assorted imported thugs, elbowed local people out of the way, including the Maltese President, to dominate proceedings. *London*'s Radio Operator Gilchrist was the only member of the British party in uniform, for Rear Admiral Oswald Cecil, last Commander British Forces Malta, and other members of the group, had been asked to wear civvies. Radio Operator Gilchrist kept his cool in the face of chaos:

> *We left the* London *around 2300 and found a large number of people milling around ashore. It was pandemonium. I was hijacked by the UK press and I particularly remember the* Daily Mail *asking me questions in such a way that they were obviously angling for something 'controversial' about what I thought to the Maltese kicking the Brits out. I had been briefed that just prior to midnight I would get the signal to lower the Union Flag. During this time there would be solemnity. I was to unclip the flag, drape it over my arm, walk to one of the Maltese dockers and shake his hand. At the same time the bugler would put the bugle to his mouth and the Maltese flag was to be hoisted to great jubilation and cheering. That is how it was supposed to be. But Colonel Gaddafi had brought 'rent-a-mob' with him.*

Gaddafi made an inflammatory speech in which he said:

> *Malta and Libya are both celebrating the kicking out of the British.*[14]

London's Radio Operator Gilchrist ignored the hostile atmosphere created by the Libyan dictator:

> *When the signal came, I did my bit, but, instead of respectful silence the British flag came down to much cheering and whooping by the Libyans. The Maltese flag went up to even louder noises. We were to remain where we were, and were presented to a passing queue of dignitaries, including Gaddafi, who didn't speak to me, and the Maltese Prime Minister, Dom Mintoff, who did, but I can't remember what he said. I think there were fireworks, music, etc etc, but the next thing I remember was returning to the ship where I was ensconced in the wardroom pantry with the Admiral and presented with a beer or two, which was heaven.*

The following day HMS *London* left, giving the ordinary people of Malta a chance to show their true feelings. The emotions displayed astonished the destroyer's sailors, including Able Seaman Eddie Cowling:

> *It was the most incredible thing. I remember thousands and thousands of people lining the ramparts, screaming, 'we love the British!' The ordinary people didn't want the Royal Navy to go. Aboard ship there were people who were crying, grown men sobbing their eyes out.*

The *London* entered refit on returning from Malta. This was completed on 8 May 1980, with the destroyer then embarking on sea trials that were interrupted by a visit to Norway and work-up at Portland, followed by exercises in home waters and a call at Brest.

The New Year saw sea time severely curtailed due to fuel restrictions and *London* finally left Portsmouth in mid-March for Exercise Spring Train, off Gibraltar.

The Last Broadside

A new generation of Type 42 destroyers armed with the Sea Dart SAM entered service with the Royal Navy from the mid-1970s and it was inevitable that the County Class, with their big crews and obsolete Seaslug, would soon be retired from service. Lieutenant Gordon Perry, who served as *London*'s Ship's Communications Officer between 1979 and 1981, has always been surprised that they got through their careers without being blown apart:

> *We were extremely lucky to have all those County Class avoid a major fire and explosion due to the Seaslug. The missile fuel was highly volatile and there were all sorts of safety precautions. The ship practised the relevant fire drills every day.*

To prolong *London*'s life in the Royal Navy it was proposed that she should be converted into a minelayer. Instead of Seaslug missiles being wheeled down the tunnel to her stern, there would have been sophisticated anti-submarine mines.

In this new role she would have patrolled up and down behind the seabed Sound Surveillance System (SOSUS) barriers, in the notorious Greenland-Iceland-UK (GIUK) Gap; the anticipated breakout point for any Soviet submarines during a war. The GIUK Gap included the notorious Denmark Strait that was so familiar to the previous *London*'s sailors during the Second World War. In the end the minelayer conversion idea was dropped and the tenth *London* was earmarked for disposal.

The Naval correspondent of Portsmouth's evening newspaper observed gloomily:

> *...she will be de-stored and then placed on the disposal list. She is more likely to go to the scrapyard than find a buyer.*[15]

It soon became clear that, like other ships of her class, *London* was an attractive proposition to a foreign fleet and was possibly destined for a new life in 'an Asian country'.[16]

Her final deployment was three months in the West Indies, where Britain still had dependent territories to safeguard. She also represented the UK in Antigua, during the island's independence ceremony. The decline in the Navy's budget continued to restrict *London*'s sea days.

...the ship must steam at economical speed, between 12 and 15 knots, throughout the deployment and aim to spend the majority of time in port rather than at sea, unless involved in exercises, fuel conservation being the name of the game.[17]

On returning to Portsmouth in mid-December 1981, the sole surviving front line British warship with two turrets was to fire the Royal Navy's last broadside. The Portsmouth-based Mary Rose Trust sent a message to *London* as she returned home, which said:

> *Greetings from everyone involved in the recovery of the* Mary Rose, *the ship that probably fired that first broadside of heavy guns more than four centuries ago... .*

The *London* signalled back:

> *From the last to the first. Very many thanks from the officers and men of HMS* London *for your kind thought.*[18]

As she progressed up the Channel, *London* was enveloped in smoke as her guns fired the historic last broadside. Her final Commanding Officer, Captain John Garnier, remarked:

> *Missiles systems are much more effective and accurate. But they do not have the same feel as the guns. With a missile you just press a button and the computer does the rest. Firing broadsides requires special teamwork.*[19]

Gordon Moores was a senior rating who had experienced that 'special team work' aboard *London*:

HMS *London* **fires the Royal Navy's final broadside.** *Topham Picturepoint/PA.*

> *Serving in the* London *was one of the happiest experiences of my career, but I hated going inside a 4.5-inch turret. Unlike today's warship guns, which are automatic, you actually had to go in there to load a 55lbs shell. You really had to mind your fingers. It was horrible and it was very loud when the guns fired. There would be about four or five people in there, carrying out a procedure that hadn't changed really since the Second World War.*

Charlie Ellis had served in *London* as a nineteen-year-old Able Seaman Gunner specializing in operating the tracking system for Seaslug, but his action station was the gun bay just below one of the turrets:

> *I helped pass the ammunition for loading in the guns. The ammunition came up on a chain rack, which kept going around and, as it reached us, we would take it off. It was not an enjoyable job, but I suppose it was better than the boredom of painting ship.*

Missing the War

The ship's crew were confident they would be handing over a fine ship, and some were convinced she still had a lot more to give the Royal Navy, had she been given the chance.

> *To cut short the working life of a good ship with a reputation as 'a good runner' such as HMS* London *is difficult to accept for those over the years who have kept her so very well... There is work to do yet to ensure that anyone who buys* LONDON *or inherits any part of her will say as I do now to you all – 'THANK YOU'.*[20]

The former HMS *London* leaves Portsmouth as PNS *Babur*. *Jonathan Eastland/AJAX.*

By the time *London*'s sailors marched behind a Royal Marine band to St Ann's Church in Portsmouth Naval Base, for the official paying off ceremony, Pakistan had been identified as the warship's purchaser. This meant that, even though *London* was still in Portsmouth when the Argentinians invaded the Falkland Islands, she did not go to war. Lieutenant Pat Lambert, one of *London*'s engineering officers, watched the naval task force being prepared for action around him:

> By then *London* only had her twin turrets fully operational, as we had fired all 32 missiles off in the Western Approaches, and on an American firing range during the Caribbean deployment. In February 1982 the Pakistani Navy took her over after the signing of a Memorandum of Understanding. In March she was officially handed over to Pakistan and, after the White Ensign had been pulled down, their sailors gradually came aboard and ours filtered off. I came in to work the morning after the Falklands thing blew up in early April and found they were ammunitioning everything in sight. I suppose if we had gone we could have played a useful role, providing Naval Gunfire support and we still had Seacat and could have flown off a helicopter. But it wasn't to be. She was commissioned as Babur on 22 April and was originally due to leave for Pakistan in May. For the Pakistanis she was quite a sophisticated ship propulsion-wise, which is why they needed our help. Their navy was going from conventional

The *Babur* photographed at Karachi in November 1985, with the Seaslug removed and superstructure extended aft. *US Naval History and Heritage Command.*

steam ship propulsion to turbines and automatic controls. I was sent for by Flag Officer Portsmouth who asked me if I could take the ship down to Pakistan. There would be me and eight senior rates seconded for the job and, two days before we sailed, they made me an acting Lieutenant Commander.

We left Portsmouth in early June, went around the back of the Isle of Wight and then headed for the north coast of France. The Pakistani navigating officer was very good and kept close to the coastline so he had plenty of landfall. I thought we would refuel at Gibraltar but they took the ship in to Toulon where she took onboard torpedoes for Pakistan's French-built submarines. One interesting incident happened when we were not far off Suez – Israeli jets buzzed us and my guys were on the upper deck waving the white flag to make sure they didn't mistakenly attack us. We were by then only helping the Pakistanis and they were doing the watch keeping themselves. When we finally reached Pakistani waters, a couple of their warships came out to welcome us in to Karachi.

Lieutenant Commander Lambert signed over the former *London* to her new owners and, after a sightseeing tour of Pakistan as guests of the Pakistani Navy, the British sailors headed home.

DLG Postscript

Ian Inskip, who served as a Lieutenant Commander in *London* during the mid-1970s, and saw action during the Falklands' War in her sister ship, HMS *Glamorgan,* is under no illusions about the County Class:

They were fine ships to serve in during flag-waving deployments in peacetime. They were beautiful-looking, almost like cruise liners, but they were designed to deal with high-flying Soviet aircraft, not sea-skimming missiles. The Counties had their Seaslug magazine above the waterline, which was a fatal flaw. If the Exocet that hit Glamorgan *off Port Stanley in June 1982 had been a few inches lower then the ship would have exploded in a most spectacular fashion.*

Despite that vulnerability, four of *London*'s County Class sister ships served on for almost another two decades under new names, and much modified, in the Chilean Navy. The former-*London* served as flagship of Pakistan's fleet until 1994 when she was decommissioned.

Notes

1. HMS *London 1961-1981 Paying-Off Newsletter.*
2. HMS *London*'s Commission Book 1963-65.
3. ibid.
4. *Evening News,* Portsmouth, 22 November 1963.
5. *Conway's All The World's Fighting Ships 1947-1982 PART 1,* edited by Robert Gardiner.
6. HMS *London 1961-1981 Paying-Off Newsletter.*
7. *Evening News,* Portsmouth, 31 October 1963.
8. HMS *London*'s Commission Book 1963-65.
9. ibid.
10. *Western Morning News,* 21 August 1965.
11. Brian Parker, in an e-mail interview with the author.
12. *Evening News,* Portsmouth, 13 March 1967.
13. *Evening News,* Portsmouth, 12 March 1968.
14. Transcript of Colonel Gaddafi's speech at Malta, 31 March 1979, Mike North Collection.
15. *Evening News,* Portsmouth, 16 September 1981.
16. *Evening News,* Portsmouth, 9 December 1981.
17. HMS *London 1961-1981 Paying-Off Newsletter.*
18. *Evening News,* Portsmouth, 9 December 1981.
19. *Evening News,* Portsmouth, 11 December 1981.
20. HMS *London,* Marine Engineering Officer's Temporary memorandum 27/81.

Chapter Eighteen

IN DANGEROUS WATERS

Bloodhound Becomes *London*

The same defence cuts that saw the tenth *London* sold to Pakistan, also threatened to curtail the construction programme that gave birth to the next warship to bear the name.

In the late 1960s the Admiralty had initiated development of a new design that would replace the Leander Class Anti-Submarine Warfare (ASW) frigate in the British fleet. This turned out to be the much larger and more capable Type 22, which would be a platform for the latest sub-hunting sonar and Sea Wolf anti-air missiles, but would not have any guns, aside from close protection cannons. Together with cheaper, less sophisticated, stopgap Type 21 frigates, the Type 22s also introduced gas turbine propulsion into the Royal Navy. As the biggest 'frigates' ever constructed for the British fleet, the Type 22s would be closer in size to light cruisers, but the term frigate had by the 1960s become currency for describing warships involved primarily in ASW, rather than being a reference to size. The reasoning behind the absence of a main gun was that the primary theatre of operations, the North Atlantic, did not require it. The standard 4.5-inch gun was most useful for bombardments in support of troops ashore but the domain of the Type 22 frigate was to be far from land. It was a radical departure, and the Type 22s revealed a clean silhouette that was too naked for some traditionalists, who regarded them as lacking martial spirit. In fact the Type 22s were far more potent than even the tenth *London* and her County Class sisters. The latter may have possessed imposing lines that included two beefy turrets, but by the 1980s they were not much more than big fat targets. War fighting capability resided in sensors and software controlling weapons with a long reach that could hit hard and fast with absolute precision.

The first Type 22 was HMS *Broadsword*, ordered in 1974 and launched the following year. But in 1981 the British government decided that, as part of its restructuring of the Royal Navy, it would curtail the Type 22 programme after the construction of

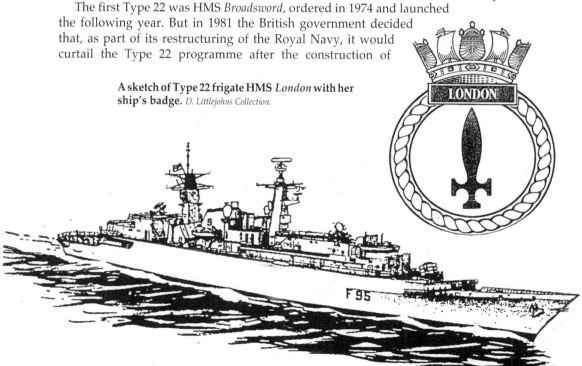

A sketch of Type 22 frigate HMS *London* with her ship's badge. *D. Littlejohns Collection.*

A Sea King Anti-Submarine warfare (ASW) helicopter lands on HMS *London*. D. Littlejohns Collection.

the seventh ship in the class, HMS *Brave*. This decision, along with the proposed sale of the carriers HMS *Hermes* and HMS *Invincible* and the scrapping of the assault ships HMS *Fearless* and HMS *Intrepid*, was greeted with outrage on both sides of the House of Commons. Thousands of British jobs depended on the Type 22 programme reaching its promised fourteen ships.

While the Thatcher government was not to be diverted from its plans to reduce the Royal Navy to a small ASW force, it was forced by the uproar to offer the construction of an eighth Type 22 frigate. On 23 February 1982 Yarrow Shipbuilders on the Clyde was awarded a £120 million contract to construct the warship. Laid down on 7 February 1983, she was launched on 27 October 1984 and would be accepted into service in February 1987. This ship was originally to be named *Bloodhound*, but the Lord Mayor of London intervened, making a special request for the warship to be named after the capital. The Lord Mayor believed the nation needed a new HMS *London* as its standard bearer on the high seas. With the sale of the previous *London*, the way was clear for the name to be released for use again.

War Comes to the Rescue
The Falklands' War of April-June 1982 saved the Royal Navy's carefully balanced fleet from destruction at the hands of its own government. It showed that carriers and amphibious assault ships still had their uses, particularly in responding to the unexpected. The performance of Type 22s *Broadsword* and *Brilliant* in the same war, confirmed what capable ships the Type 22s were, even though success came not in destroying submarines, but in killing enemy aircraft. On the picket line that acted as the British task force's only means of early warning and front line defence, the Type 22s were teamed with Type 42 destroyers, the so-called '42-22 Combo'. While the destroyers' Sea Dart missile had a long reach, they had some blind spots that the shorter range Sea Wolfs of the Type 22s could cover. In the wake of the conflict, the number of Type 22s to be constructed for the Royal Navy was restored to fourteen.

The Type 22s came in three batches – four Batch 1s, six Batch 2s and four Batch 3s. The Type 22 Batch 1s had hull-mounted sonar, but the Batch 2s were fitted with more effective bow mounted short-range sonar, requiring a raked bow. Batch 1 Type 22s

Captain Doug Littlejohns supervising a RAS from one of *London*'s bridge wings.
D. Littlejohns Collection.

The towering sides of *London* were deeply impressive to a submariner. *Nigel Andrews/active vision.*

could only carry one Lynx anti-surface/submarine helicopter while the Batch 2s could carry two Lynxes or one Sea King. In anti-submarine operations the Lynx lacks the endurance of the Sea King and does not carry as many ASW sensors or weapons. However, the powerful surface search radar of the Lynx, combined with Sea Skua anti-surface missiles, had proved formidable during the Falklands' War.

Aside from a larger hangar and flight deck the Batch 2s were also longer, improving endurance and sea keeping (the Batch 1s tended to roll badly in rough weather). The increase in hull size also allowed the Batch 2s to carry towed array sonar trailed from the stern of the ship and carry more surveillance equipment and improved command and control systems. The Batch 2's American-origin Outboard Electronic Intelligence (ELINT) system allowed an amazing array of intelligence gathering and enabled superb task group integration. A leaflet handed to visitors to *London* explained how everything merged:

> *The ship's communications facilities and sensors are part of an integrated system which feeds a centralised computer system. The heart of the ship is the Operations Room where radars, sonars and other sensors feed their information into the computer system, which displays this data in a rapid and informative format to enable the Command to evaluate and counter any threats to the ship.*

The eleventh *London* was 485 feet long, with a beam of forty-eight feet. Her displacement was 4,800 tons (full load) and her crew varied between 265 and 286, depending on her mission (flagship duties and the need for additional surveillance specialists could boost numbers). She was armed with two Sea Wolf launchers – one on the hangar roof and the other on the forecastle – both containing six missiles, but capable of reloads. She also had four Exocets in single-use launchers on her forecastle below the Sea Wolf. To deal with aircraft and small boats at close range she was armed with two 30-mm twin barrel and two 20-mm single barrel gun systems. The Type 22 *London* also had two triple launchers for Stingray lightweight anti-submarine homing torpedoes, one on either side amidships. Powered by Rolls-Royce Olympus gas turbines for high speed and Rolls-Royce Tynes for cruising, frigate *London* could manage a top speed of thirty knots. Her range at eighteen knots was 4,000 miles .

Improvements inspired by experience in the Falklands were applied throughout the Type 22s. These included dualling of water mains that fed fire hoses, use of non-toxic coatings for wiring and thick curtains to be drawn across compartment boundaries to prevent the spread of smoke. The four Batch 3 Type 22s received a Goalkeeper Close-in Weapons System (an air defence Gatling gun) and the standard British naval 4.5-inch gun for shore bombardment, which also has an anti-air capability. These powerful ships, which are still in service today, are also armed with Harpoon anti-shipping missiles.

Shadow Boxing With the Soviets

During patrols as commanding officer of the nuclear-powered attack submarine HMS *Sceptre*, Captain Doug Littlejohns had stalked surface warships, but in late 1987 he was in the novel position of being appointed to command a 'skimmer'. He took over command of HMS *London* from Captain Robert Fisher, the man who had brought her out of build, through sea trials and commissioned her into service.

London's new Commanding Officer found her an imposing sight. Having been used to submarines, which have most of their bulk beneath the sea, the towering sides of a Type 22 frigate were a shock.

> *I remember driving down the jetty to find the ship and being absolutely overwhelmed by the length and height of her compared with my last command. There wasn't much difference in terms of tonnage, but the* London *seemed huge by comparison.*

The frigate would need every inch of space in her large hull to pack in sensors capable of handling an increasingly tricky opponent.

Soviet patrol craft shadow HMS *London* in the Baltic. *D. Littlejohns Collection.*

While the Soviet fleet of the Second World War had been an unadventurous, largely coastal fleet, content to let the Royal Navy fight the Germans almost on its own during the Arctic convoys, after four decades of Cold War it had become a truly global force. By the late 1980s the quality of Soviet submarines and surface warships had improved greatly and Western navies, accustomed to thinking they were light years ahead of the Soviets, received some nasty shocks. The Soviets' new Alfa Class hunter-killer submarines were discovered to be incredibly fast (capable of forty-five knots underwater) and, with a revolutionary titanium hull, could dive to 800 metres. Stealthy Kilo Class diesel-electric submarines had also made their debut. New surface warships included Kiev Class helicopter/jump jet carriers; Kirov and Slava class missile cruisers; Sovremenny anti-shipping and Udaloy anti-submarine destroyers. New ships were also given the ability to reload their missile launchers, enabling sustained combat. It was clear proof that quality was getting the upper hand in the Soviet fleet. Although its surface warships were still, at first glance, crude compared with Western vessels, they were functional, tough and fitted with some superb weapons systems. Typhoon Class ballistic missiles submarines – the world's largest at 25,000 tons displacement – which could hit targets with nuclear missiles from the security of 'bastions' under the ice in the Barents Sea, had also been built by the Soviet Union. New Akula Class attack boats were coming into service and their first big aircraft carrier, capable of launching fixed wing jet fighters, was about to commission with more like her to follow. The Soviet Navy was divided into four main fleets – the Northern (Red Banner) Fleet, based in the Arctic north; the Baltic Fleet, with main bases at Kronstadt in the Gulf of Finland and Kaliningrad in the south; the Black Sea Fleet, centred on Sevastopol and the Pacific Fleet, concentrated at Vladivostock and in the Kamchatka Peninsula.

With no comparable NATO fleet to rival them in the Baltic, the Soviets considered it to be their own private lake. Sophisticated warships like HMS *London* were constructed as a response to the Soviets' burgeoning naval might and Captain Littlejohns was determined that HMS *London* would stir things up. In January 1988, almost immediately after work-up had finished, *London* set sail for the Baltic to give her powerful sensors a thorough test-drive against the units of the Soviet Navy:

We went through the gap between Denmark and Sweden and into the Baltic at night, disguising the warship's lights to make her look like a merchant vessel. The following morning we were somewhere off the Baltic States, which were at that time a part of the Soviet Union. Our main purpose was hanging around sucking in useful intelligence on our sensors, making sure they all operated up to spec. Once they found us in their backyard the Soviets reacted immediately by sending over aircraft and helicopters to take a look at us. Then a Matka patrol vessel turned up as our permanent shadow. We had some fun and games with him. If we went fast then he was OK, but if we went slowly he would coke his engines up after an hour or so and would have to do a high speed run to burn it off. On a Sunday morning we ran up the church service pennant for a bit of a spoof on the Matka. Beforehand we had made up this mini coffin out of plywood or something and we brought this up. I played the part of the padre and we sang a few hymns during a 'funeral service' before heaving this 'coffin' over the side.

Inside it had a copy of Hunt for Red October, *one of around a dozen or so in a Russian language edition, given to me by the author Tom Clancy, who was a pal of mine. Also in the 'coffin' was a bottle of whisky and copies of* Penthouse *and* Playboy, *plus a little note saying something like 'haven't you got anything better to do?' We had made sure that it would float*

The *London* alongside the Second World War cruiser *Belfast* on the Thames. *D. Littlejohns Collection.*

and, as anticipated, the Russian couldn't resist picking it up. No doubt he had a shock, but I am sure he enjoyed our decadent capitalist gifts. He was certainly less unfriendly thereafter.

The *London*'s sailors also sought to tease their Cold War shadows by creating exotic looking weapons systems. Tim Allen, who had served on the previous *London* as junior rating and was one of the Type 22's petty officers, found it to be a most enjoyable sport:

> *While we were in the Baltic the First Lieutenent organised a competition to design the best, biggest and weirdest looking torpedo, to get the Soviets talking and see what their reaction would be. Some of our seamen put half a dozen dustbins together with lids added here and there, plus the biggest tail fin in the world and painted it various camouflage colours. I think it was about 25-feet long in the end and really looked good situated by our real torpedoes. The Soviets were very impressed, as we could see them taking lots of photographs of the 'torpedo'.*

One night while testing the sonar, *London* detected a Russian submarine lurking nearby. Captain Littlejohns decided to see how Soviet submariners performed under pressure:

> *It couldn't have been anything else down there but a Russian boat shadowing us. So, we went back and forth, getting some very nice traces that revealed it to be a Foxtrot Class diesel-electric submarine. Then we decided to sit above him until he ran out of air or he needed to come up and re-charge his battery. Next morning, when he did surface, we circled him and flashed: 'Good morning, hope you are having a good day'.*

The Tanker War

On returning from the Baltic deployment in April 1988 the ship visited London. It was the second time she had been there, but her first as a fully commissioned warship. Captain

Littlejohns appeared to court controversy by enacting a right traditionally enjoyed by Commanding Officers of Royal Navy warships named *London*.

> *We fired a gun salute to the Lord Mayor much to the chagrin of the Keeper of the Tower of London who claimed that we were in breach of a document signed by Oliver Cromwell that disallowed ships of the Royal Navy from uncovering their guns in the Pool of London. There were letters to the newspapers for months afterwards, but I had taken care to obtain permission to fire the salute from the Ministry of Defence well in advance.*

The next deployment for HMS *London* would take her to the Gulf at a time of high tension, for the war between Iran and Iraq was claiming lives at sea.

The conflict had started in 1980 and partly revolved around who should have sovereignty over the Shatt al Arab waterway at the head of the Gulf and possession of the oil-rich Iranian province of Khuzestan.

Iraqi dictator Saddam Hussein had assumed power in Baghdad in 1979 not long after Islamic revolutionaries led by the Ayatollah Khomeini had overthrown the Shah of Iran. In addition to coveting Iran's territory and natural resources, Saddam viewed the Iranians as potential rivals for dominance over the Gulf. Iraq invaded Iran on 22 September 1980 and the fighting soon bogged down into trench warfare. Fearful of the virulently anti-Western stance of the Iranians, the USA and other nations, including Britain, joined the Gulf States in supporting Saddam against Iran, supplying arms and covert intelligence. The Soviets, engaged in their own war against Islamic fundamentalist fighters in Afghanistan, also supported Saddam. The Chinese supplied Iran with weapons, including Silkworm anti-shipping missiles, while the Iraqis were able to obtain Exocet missiles from France, together with Etendard and Mirage strike jets to carry them. Each side was, therefore, well equipped to threaten maritime trade and this '...forced the navies of many nations to escort and protect their merchant ships in the area – especially oil tankers, regarded as fair game by both belligerents.'[1]

The USA provided the biggest naval presence in the area particularly after March 1987 when Kuwaiti tankers were allowed to reflag as American ships. The Royal Navy, which had withdrawn from 'east of Suez' at the end of the 1960s, found itself dragged back there, mounting regular patrols in the Gulf to safeguard UK-flagged merchant vessels, including Kuwaiti.

While Iran was thought to pose the main threat to shipping, it was the Iraqis who caused the first naval casualties. On 17 May 1987 one of their Mirage F1s attacked the frigate USS *Stark* after mistaking her for a tanker. Two Exocets hit the American warship, killing thirty-seven sailors. Saddam apologized profusely for the mistake, but some wondered if it was entirely genuine. Over the next few months several tankers struck mines off Kuwait, while in the Straits of Hormuz – the narrow mouth of the Gulf – the Iranians were stopping and searching merchant ships to make sure they were not taking goods to Iraq or exporting its oil. Silkworm missiles launchers were meanwhile being positioned by the Iranians to dominate the waters of Hormuz, while in the north Revolutionary Guard fast attack craft conducted interference patrols.

In late July the US Navy provided an escort of three warships for two Kuwaiti tankers that had been reflagged as American. One of them, the *Bridgeton*, hit a mine but did not sink. By September the Americans were running aggressive search and destroy missions with their naval helicopters against minelaying vessels and attacking abandoned oil platforms,

The hot zone. *Crowter/D. Littlejohns Collection.*

suspected of hosting Iranian surveillance teams. On 14 April 1988, the US Navy frigate *Samuel B. Roberts* hit a mine while on patrol off Bahrain and was almost broken in two. There were no fatalities and somehow the warship limped to safe harbour. America hit back four days later with Operation Praying Mantis, attacking two abandoned oil platforms thought to be hiding Iranian forces. The platforms were destroyed and, in the battle that developed, the Iranians lost a missile boat and a frigate, with another frigate badly damaged.

By the summer tension was higher than ever and the Americans expected the Iranians to launch reprisal attacks against United States Navy warships at any moment. On 3 July the missile cruiser USS *Vincennes* was embroiled in a series of running battles with Iranian fast boats. In the middle of fending off the Iranian vessels her radar appeared to indicate an air attack was inbound. The American warship shot down an aircraft, but it turned out

London **rides shotgun during the Tanker War.** *D. Littlejohns Collection.*

SHIPS ESCORTED DURING THE MONTH OF AUGUST

SHIPS NAME	BWT TONNAGE	IN/OUT
Sandgate	27,401	OUT
British Skill	127,778	OUT
Helikon	17,651	OUT
Tokyo Bridge	24,037	OUT
World Hitachi Zosen	268,904	IN
Tidespring (RFA)	18,900	IN
Mohdi	26,139	IN
Tidespring (RFA)	18,900	OUT
B.T. Trader	319,294	IN
B.P. Vision	89,735	IN
Havpil	11,470	OUT
Isomeria	47,954	IN
Red Seagull	486,258	IN
Esso Kawasaki	302,577	IN
Esso Pacific	500,268	IN
Chevron Edinburgh	268,333	IN
Providence Bay	34,477	OUT
World Brasilia	283,761	OUT
Coulonge	31,275	IN
Australian Star	16,114	OUT
Chilham Castle	27,401	IN
Hermod	18,165	IN
Olna (RFA)	25,100	OUT
Eriskay	229,936	IN
Helikon	17,651	OUT
Leonia	317,996	OUT
Eastern Strength	267,577	OUT
Olna (RFA)	25,100	IN
Chevron Edinburgh	268,333	OUT
Zenatia	57,741	OUT
Burnpac Bahamas	254,891	OUT
Australian Star	16,114	IN
Middleton (HMS)	725	IN
Olna (RFA)	25,100	OUT
Chevron South America	413,158	IN
British Reliance	269,757	OUT
B.T. Trader	319,294	OUT
Havpil	11,470	IN
Helikon	17,651	IN
British Respect	277,747	
Mandama	19,114	

OUT = Persian Gulf to Gulf of Oman.
IN = Gulf of Oman to Persian Gulf.

The weights of R.F.A's and warships escorted are included in this list, but they are not counted to the total tonnage escorted.

TOTAL TONNAGE without R.F.A's and warships = 5,615,422.
TOTAL TONNAGE with R.F.A's and warships = 5,729,247.

The *London*'s Tanker War escort total for August 1988, as published in the frigate's newsletter. *Crowter/D. Littlejohns Collection.*

to be an Iranian airliner carrying 290 people, all of whom died. The Iranians came close to declaring all out war on America.

This was the intense operational environment into which HMS *London* sailed when ordered to the Gulf at short notice in the summer of 1988. Her original deployment date had been mid-July but everything changed after Captain Littlejohns received an urgent phone call from Fleet Headquarters on 12 June.

One of our Leander Class frigates in the Gulf had a problem with fuel contamination. So the London was required to go in ten days time instead of a month and we had to recall everyone back from leave. The ship had not completed her pre-deployment work-up and was supposed to go back to Portland for damage control training. Instead we ended up sailing with some of the Frigate Squadron people aboard and did our training during the journey to the Middle East. We did about 28 knots all the way stopping in Gibraltar for five hours for our first refuelling. We refuelled at the French naval base in Djibouti and then at sea off the Straits of Hormuz. Initially we patrolled the Indian Ocean just off the straits and carried out some exercises with the Americans, 'sinking' the carrier USS Forrestal with our Exocets, which was quite rewarding.

The *London* was temporarily diverted to Pakistan to visit Karachi to support a UK arms sales drive. Captain Littlejohns recognized a familiar silhouette across the harbour.

We parked the ship just across from the previous London in her new guise as the Babur and flagship of the Pakistani Navy. After leaving Karachi we had to send a signal back to the UK saying our ship was non-operational because 75 per cent of the crew were suffering from conjunctivitis due to Karachi's polluted air. We went to Muscat in Oman to recover and a couple of days later the ship was available for escorting merchant ships into the Gulf.

There were two congregation points for tankers – one on the Arabian Sea side of the Straits of Hormuz for tankers going in, and another off the United Arab Emirates for vessels coming out. After making radio contact with the tankers, Captain Littlejohns took *London* in close to the gigantic ships, some of them twenty times bigger than the frigate, and attempted to arrange them in a manageable group.

The London would go up and down among them like some sort of sheep dog, getting them

HMS *London*'s ships company, 1988. *D. Littlejohns Collection.*

in order for the transit. As Western warships took the tankers through the Straits the Iranians would sometimes send their warships out to sail up and down parallel to the convoy. I remember one of their British-built frigates actually trained his guns on London. *I didn't take too kindly to that and got on the radio to tell him: 'I would really prefer you don't do that!' Fortunately he turned his guns away.* London *also made a few air contacts on her radar and we also had to watch out for mines. Then we had an incident where we had a report of some Iranian missile boats leaving an island base, in an apparent attack formation. I scrambled both of my Lynxes with Sea Skua anti-shipping missiles and the Airborne Early Warning Sea King from the Royal Fleet Auxiliary ship that was with us. The Sea King reported a very good surface picture back to* London. *We saw the Iranians coming at us, but turned around and went back in to their base, so we were able to stand down.*

On 20 August 1988 a ceasefire came into effect; a grudging truce between the two exhausted combatants forced on them by the international community appalled at the wanton, senseless destruction and 1,000,000 casualties. The Iraqis had achieved nothing, gaining a useless portion of Iranian territory. They were saddled with huge debts, owing Kuwait alone US $14 billion, which would become a source of conflict in itself. As the Gulf settled down into an uneasy peace, that would eventually prove to have been only an interlude, HMS

London's deployment neared its end, but her departure for home was delayed when the Type 42 destroyer HMS *Southampton* received near catastrophic damage during a collision with an oil tanker. *London* was required to fill the patrol gap until a relief warship arrived from the UK. Captain Littlejohns recalled:

> Southampton's *accident meant we had to stay out there for several more weeks, and in the end had the dubious honour of holding the Royal Navy record for tonnage of ships escorted in the Gulf. I cannot remember the exact number but it was over five million tons.*

Captain Littlejohns brought *London* home to the UK and, with an important desk job in the Ministry of Defence on the horizon, then made the most of taking her on a major exercise in the north Norwegian Sea.

> *I would much rather have stayed in command of* London, *because I knew she was going up to the Barents Sea in the summer of 1989. I wanted to take her up there because I had been up in northern waters in a submarine and it would have been excellent to go back and do it in a frigate. But, it wasn't to be.*

Bold as Brass in the Barents

The *London*'s new Commanding Officer was another submariner who had commanded the nuclear-powered attack submarine HMS *Spartan* during the Falklands' War. Captain James Taylor stalked Argentinian warships during that and, like Doug Littlejohns before him, was now expected to perform the transition from 'poacher' to 'gamekeeper' seamlessly:

> *Both ship and crew were fantastic. The Batch 2 Type 22s had enormous legs, thanks to their tremendous fuel capacity. They were designed to operate on their own 100s of miles from the nearest friendly force.*

And in the summer of 1989 that is exactly what HMS *London* was tasked with doing, back in waters so familiar to her namesake in the Second World War. Taking a Royal Fleet Auxiliary tanker with her, she headed into the Barents Sea as the Northern Fleet was holding a series of massive exercises that included live firings. *London*'s job was to watch, take notes and make off with anything that came to hand. The frigate was operating in just the role she was designed for and Captain Taylor was one of thirty Russian speakers aboard, including specially assigned communications ratings.

> *We all had a useful working knowledge of Russian, or better, and it helped us to understand what was happening around us. The* London *got in among the missile and gunnery shoots that the Russians were conducting, checking information on their sensors and telemetry. Every now and then the duty marking vessel assigned to keep tabs on us by the Northern Fleet would come over the horizon and stick really quite close. You had to make sure the rules of the road in international waters were on your side and you always had your camera at the ready and they, of course, had theirs pointed at us. By keeping such close proximity with them you risked giving away a little about how you operated but you gained a lot more information on them.*

Yorkie Cunningham, who was one of *London*'s sonar senior ratings at the time, recalled that the ship was lucky enough to catch sight of a Typhoon ballistic missile submarine:

> *We were on the 12-mile limit, just outside their territorial waters, and we had just witnessed their submarines firing missiles, when this Typhoon came up on the roof, surfacing four or five miles away from us while still in international waters. We sent one of our helicopters after her and I believe we got the first head-on photo of a Typhoon on the surface.*

In the Cold War, intelligence photographs were the equivalent of notching up a kill but *London* also tried to pick up a piece of Soviet hardware. James Taylor recalled how he nearly pulled off a real coup:

> London *got into a torpedo-firing exercise by a Victor-class SSN and secured a line on the latest Soviet heavyweight torpedo. This was very high on the intelligence shopping-list I had been given, but my instructions were only to remove hardware if I was unobserved. A Soviet destroyer was sitting almost alongside me. While he invited me to let the torpedo go, and I feigned deafness, our ship's diver gave it a thorough examination under the water before we slipped the line.*[2]

The Beginning of the End

Returning to Devonport, HMS *London*'s specialists provided some debriefings on what they had seen in the Barents Sea and in the autumn the frigate sailed for an exercise in the North Atlantic that enabled her to fire one of her Exocets. In November *London* visited the German naval port of Wilhelmshaven, giving some of her crew a ringside seat for one of history's decisive moments. Mikhail Gorbachev – president of the USSR and Secretary General of its Communist part since 1985 – had realized that the dream of a healthy, modern Soviet Union was being suffocated by the vast amount of resources devoted to its gigantic war machine. The USSR's economy simply could not afford to continue the arms race with the West. The cuts in nuclear missiles agreed with the Americans in the late 1980s were the exterior face of Gorbachev's programme of reform, which the Soviet leader wanted satellite states, within the Soviet orbit, to embrace as well. But, while Gorbachev saw the USSR and its Eastern Bloc subjects continuing in a new, modernized Communist harmony, they interpreted it as the beginning of the end. In the winter of 1989 the Warsaw Pact began to fall apart, with Hungary rejecting Communism first, Czechoslovakia finally throwing off the shackles, and East Germany coming apart at the seams.

For East Germany, it culminated in the Berlin Wall being torn down on 9 November and the Cold War warriors of HMS *London* were there to see it happen. Captain Taylor and five other officers from the frigate jumped on a train to Berlin so they could witness history in the making.

> *It was the most amazing moment. There were people having a go at it with implements and others coming through holes in it. One man was having a go at the Wall with a sledgehammer and shouting: 'I hate this bloody thing!' It was the best party any of us had ever been to and I even got to take away a piece of the Wall.*

No one caught up in the joy of celebrating the beginning of the end for the Cold War could have guessed that, in little more than a year, HMS *London* would be back in the Gulf, leading a task group of British warships in a hotly contested war of liberation.

Notes

1. Stephen Howarth, *To Shining Sea*.
2. Jim Ring, *We Come Unseen*.

Chapter Nineteen

SADDAM'S APOCALYPSE

Debts of Dishonour

Iraqi dictator Saddam Hussein was, by the summer of 1990, increasingly bellicose towards his Arab neighbours. He was very angry about the Gulf States demanding repayment of debts that he felt had been incurred fighting a war to protect them from Iran.

They were adding insult to injury by producing too much oil, which was depressing the cost per barrel and therefore undermining Iraq's revenues. The Kuwaitis in particular were a source of annoyance, still pursuing their US $14 billion loan and, according to Saddam, also stealing oil from Iraq.

The Rumaila oilfield is bisected at its southern end by the Kuwait-Iraq border, and the Iraqis claimed the Kuwaitis were drawing off oil from the wrong side. By July 1990 the Iraqis were demanding drilling rights within the Kuwaiti portion of the Rumaila and suggesting the sheikhdom should 'loan' them more money to help write off their international debts. Saddam hinted that Iraq might resort to military action if something was not done to ease its economic misery, such as a decrease in oil production leading to a rise in the price per barrel.

Iraq had, several times in the past, claimed that Kuwait was a 'lost province', threatening to reclaim it by invasion in 1961. Only the timely arrival of Royal Marines and Army troops, together with armour and warships deterred the Iraqis from seizing the former British protectorate. Nearly thirty years later it would take a lot more than a few rapid-reaction units to stop Iraq if it decided to occupy oilfields in neighbouring Kuwait. Even after the huge casualties of the recently ended war with Iran, and despite being saddled with major debts, the Iraqis maintained the fourth biggest army in the world. It '...ranked in size behind only those of China, the Soviet Union, and Vietnam. (The United States ranked seventh)'. [1]

With 900,000 soldiers in sixty-three divisions, Iraq plainly had a much bigger military than it needed for self defence purposes. While the Iraqi command structure was rigid and its logistics support poor, it had some of the most sophisticated weapons the world could offer, including Silkworm and Exocet anti-shipping missiles, together with Mig29, Su24

HMS *London*, followed by Royal Navy minehunters and in company with a US Navy battleship, 'Up-Threat' in the Gulf War. *C. Craig Collection.*

and Mirage F1 strike jets. The Iraqi Navy consisted of a training frigate, ten Soviet-built Osa fast attack craft (armed with Styx anti-shipping missiles) and ten other patrol vessels. Four frigates were under construction at an Italian shipyard. The Iraqis also had eight coastal minesweepers, seven military transport vessels, an unknown number of auxiliary patrol vessels and hovercraft used by their special forces. Western intelligence experts were concerned that the Iraqis might even have a midget submarine in service. While Saddam's fleet was small, the Osas did pose a potent threat if handled properly and the shore-based Silkworm batteries were capable of sinking a warship with one hit. The training frigate could also be adapted to carry missiles. More frightening than any conventional weapons the Iraqis possessed, was their propensity for using gas – employed several times in the 1980s against the Iranians and the Kurds – and their fascination with weapons of mass destruction. In March 1989, British law enforcement officers had arrested a group of people trying to smuggle nuclear weapons' triggers to Iraq. The following month the world learned about Iraq's programme to construct a 'supergun' capable of firing artillery shells into Israel, after Britain impounded massive metal tubes thought to be parts of its barrel.

A Multi-national Armada
When three of Saddam's elite Republican Guard Divisions swept across the border and into Kuwait City on 2 August 1990, the world was stunned. The build-up of Iraqi troops close to Kuwait had been no secret but the most American and British intelligence experts expected was a temporary incursion into the oilfields. Now it looked as if Iraqi forces might thunder on into Saudi Arabia, which was another potential target for revenge, the Saudis having joined Kuwait in raising oil production while requiring Iraq to repay war loans. If Saddam occupied Saudi Arabia he would control '...40 per cent of the world's known oil reserves'. [2] The Iraqi dictator would be able to hold the world to ransom, with the power to destroy the global economy by turning off the taps if his demands were not met.

On 3 August the United Nations Security Council unanimously passed Resolution 660, which demanded a complete Iraqi withdrawal from Kuwait. Three days later the UN Security Council passed Resolution 661, imposing sanctions and a trade embargo on Iraq. This prohibited the import and export of all goods except medical and humanitarian supplies and would need a sizeable naval force to enforce it. The embargo began on 31 August, with an American warship intercepting an Iraqi tanker. In the meantime, under the banner Operation Desert Shield, troops and warplanes from America, Europe and other Arab countries began pouring into Saudi Arabia to ensure Saddam's divisions could not go any further. At sea there was an equally massive build-up of striking power, with the Americans and their allies assembling the mightiest fleet since the Second World War. It would ultimately embrace half a dozen Carrier Battle Groups, two task groups centred on the battleships USS *Missouri* and USS *Wisconsin*, more than a dozen nuclear-powered attack submarines and four Amphibious Ready Groups carrying thousands of US Marines. A total of eighteen nations made up the Allied naval coalition, including Australia, Canada, France, Italy and the Netherlands. Even Kuwait managed to contribute two missile-armed attack craft that had escaped from the Iraqis. Eight Kuwaiti missile boats, armed with Exocets had been captured, but it was doubtful their missile systems were working. The next most powerful force committed to the Allied armada in the Gulf was the Royal Navy's. It would have contributed more ships but was hamstrung by on going post-Cold War 'peace dividend' defence cuts. The carrier *Ark Royal* and assault ship *Fearless* were not sent because the money needed to prepare them specifically for the Gulf combat environment was not made available. At the time it was alleged that *Ark Royal* could not go into the Gulf because the UK government wasn't prepared to pay for the extra engines her Sea Harriers would have needed to operate in the demanding desert environment. *Fearless* lacked full Nuclear Biological and Chemical Defence protection, despite having just come out of a refit. In the end the *Ark Royal* headed a task group in the Eastern Mediterranean mainly to watch Libya, an ally of Iraq. This Royal Navy group was joined by American warships that could fire Tomahawk land attack missiles against targets in Iraq.[3] Luckily for the Royal Navy task group in the Gulf, the Americans willingly provided detection and long-range interceptor assets to counter the Iraqi air threat. Even so, the British contribution was significant, for a total of thirty-one Royal Navy vessels were deployed on front line service between August 1990 and April 1991.

Gulf Flagship

With deployment of a carrier ruled out, another Gulf flagship had to be found and HMS *London* was selected for the job. Arriving in theatre under the command of Captain Iain Henderson in early October 1990, she had been equipped with additional high-powered communications kit to enable her to carry out the role.

The *London* initially provided a command base afloat for Senior Naval Officer Middle East (SNOME) Commodore Paul Haddacks and his staff, but on 3 December Commodore Chris Craig arrived to take charge. Like his predecessor, Craig was a Captain promoted specifically, and temporarily, for the job of commanding the task group. He found the *London* by no means ideal but in the end Commodore Craig felt the frigate did an excellent job:

> The London *was emphatically cramped as a flagship – often my flag staff officers out-numbered the ship's own. We ended up with 26 ships in the task group at the height of the war and 18 helicopters to boot. It wasn't easy exercising effective Flag Command of such a large task group from a Type 22, but in true British style we improvised. We did not have a dedicated command centre or accommodation so we made do with the ship's Anti-Submarine Warfare command facilities.*

Commodore Chris Craig, left, and Captain Iain Henderson, Commanding Officer of HMS *London*, inspect their surroundings. *Toby Elliott Collection.*

The *London*'s ASW personnel were also attached to the Commodore's staff:

> So, it was just as well that at that time there was no submarine threat. There is no doubt, with hindsight, if one had known we would be involved in such a major conflict that the MoD would have been looking straight away at deploying a two star admiral with his command staff in a carrier. In reality though, the situation was not so clear-cut and perhaps the people in the UK didn't realise it would end up as a full shooting war. What I thought was brave of the Ministry of Defence and Fleet HQ at Northwood was to persevere with the staff and command structure already in place, to stick with the people who had a handle on the situation. To have effected a command shift in mid-conflict would have been more destructive...out-weighing the benefits of the carrier being sent. A new task group commander would have found re-establishing good links with the Americans very difficult to manage when the missiles were flying.

Chief Petty Officer Yorkie Cunningham and Petty Officer Bob Burton led the two watches of *London*'s sonar ratings who ran situation maps for the Commodore. Bob Burton explained:

> We had acetate markers to denote each ship, which we moved around when required. It was all so the Boss would be able to tell at a glance what was happening and it was accurate to within four hours.

Each watch was composed of a senior rating, a Leading Seaman and an Able Seaman working six hours on, six hours off. As a sonar specialist it was not what Chief Petty Officer Cunningham had trained for, but he soon adapted to his demanding new role while ensuring *London* could still hunt submarines:

> When we shut down the Sound Room, we put boards in front of all the sets, but we used to start the sonar up once a week to make sure it was all still working and ready to go if needed. I believe there might have been a potentially unfriendly submarine somewhere in the Gulf but it never left harbour. Our job basically consisted of collating information. Every warship and every merchant ship in the operational area had to be kept track of – there were 130 warships alone. We also had to sift the Commodore's signals – he could get 800 in a day and we had to prioritise them.

Warning Shots

Not long after arriving in theatre, *London* was in the Gulf of Oman, sailing in company with Batch 1 Type 22 frigate HMS *Battleaxe*, the Australian destroyer HMAS *Adelaide* and US Navy destroyer USS *Reasoner*, when a vessel that might be attempting to run the embargo was detected.

The Iraqi-registered ship was asked to stop and allow a search but refused, forcing the frigates to fire bofors cannon shots across her bows. Two six-strong Royal Marines' boarding teams then rapid-roped down onto the deck of the ship from hovering Lynx helicopters. They met no resistance and the vessel was handed over to the custody of US naval forces for a more detailed search. Keeping everyone sharp during boardings was the thought that Saddam might have ordered suicide squads to secrete themselves in merchant ships, ready to blow them apart once Allied troops were aboard, damaging any nearby warship in the process.

In his cramped HQ aboard HMS *London*, Commodore Craig also contemplated the potential conventional threat:

> Iraqi combat forces were hardened by nearly ten years of war and possessed sophisticated air defence and a proven electronic warfare capability. Their anti-ship assets included 50 Mirage

fighter-bombers, 400 air-launched Exocets and 7 shore launchers with 50 Silkworm anti-ship missiles...They also owned a miscellany of sea mines.[4]

By December 1990 the Royal Navy task group in the Gulf included HMS *London* and Type 22 (Batch 1) sister ship HMS *Brazen*, Type 42 air-defence destroyers HMS *Gloucester* and HMS *Cardiff*, three Hunt Class mine-hunters and their headquarters ship HMS *Herald*, the Royal Fleet Auxiliary battle damage repair ship RFA *Diligence*, forward casualty reception vessel RFA *Argus*, replenishment tanker *Olna* and the supply ship *Fort Grange*.

Captain Henderson was keen for HMS *London* to test her Sea Wolf system so, in late December, the frigate headed for the more open waters of the Arabian Sea. But, before the firing could be conducted, on the last day of 1990, HMS *London* joined forces with an American warship and the Australian frigate HMAS *Sydney* to intercept the 36,000 tonnes Iraqi oil tanker *Ain Zalah*. While *Sydney* took the role of coordinator, *London* and the American warship moved in close, with helicopters from all three dropping boarding teams onto the tanker. They met no opposition and a search failed to reveal contraband, so the *Ain Zalah* was allowed to go on her way to Iraq. The Iraqi government later claimed the tanker's crew had been roughed up and the vessel damaged in what amounted to 'an act of piracy'.[5] The allegations were unfounded. The interceptions went on and by year's end the coalition's maritime force had challenged 2,000 vessels, with twenty-five boardings, but with with just one ship diverted for further searches.

Turning up the Heat

After firing her missiles and exercising with American naval units, *London* passed back through the Straits of Hormuz on New Year's Day 1991. It looked increasingly unlikely that diplomatic pressure and the embargo alone were going to force the Iraqis out of Kuwait, so preparations for the biggest military offensive since the Second World War were well advanced.

With time to find a peaceful solution rapidly running out, the coalition navies got down to rehearsing for the worst. For the British that meant participation in Exercise Deep Heat, held between 6 and 8 January, which was a wide-ranging combat rehearsal. It included 'rescuing' a crippled vessel under air attack. This required some dramatic manoeuvres from participating warships, as a newspaper correspondent with the task group reported:

> *Then it was* Gloucester's *turn. Her bridge crew shouted frantic orders as they imagined the low-flying Jaguars were Iraqi Migs or Mirages. The mock strafing brought back memories of 'bomb alley' in San Carlos water during the 1982 Falklands War.*[6]

Deep Heat culminated in a firepower demonstration off Qatar witnessed by British Prime Minister John Major, who was visiting British forces in the Gulf. In his latest dispatch from

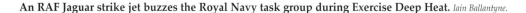

An RAF Jaguar strike jet buzzes the Royal Navy task group during Exercise Deep Heat. *Iain Ballantyne.*

The Type 42 destroyer HMS *Gloucester*, riding in close company with HMS *London*, just north of Qatar, a few days before the Allied air offensive began. Her Lynx helicopter (sitting on her flight deck) would play a key role in wiping out the Iraqi Navy. *Iain Ballantyne.*

the task group, the newspaper correspondent aboard *Gloucester* set the events at sea within the broader context:

> *Prime Minister John Major is due on board Devonport frigate HMS* London *today as his four day visit to the Gulf crisis zone reaches its climax. He will see the 14 vessels of Britain's Middle East naval task force preparing for war if tomorrow's last ditch talks between US Secretary of State James Baker and Iraqi Foreign Minister Tariq Aziz fail.*[7]

During a brief visit to the *Gloucester* Commodore Craig told the same correspondent:

> *We have got to consider the worst case...That could mean conflict with the nation of Iraq.*[8]

The British naval commander went on to say that he hoped the collective experience of his ships' commanders, many of them blooded in the Falklands' War would stand the ships under his command in good stead. After the Commodore returned to the *London*, Deep Heat's final day of action started.

Royal Air Force Jaguar jets buzzed the task group, weaving in and out of the ships at wave top height as they dropped flares, then the 4.5-inch guns of *Gloucester* and *Cardiff* boomed and the cannons of *London* and *Brazen* spat fire, wreathing the ships in smoke.

On 11 January the British task group went to Air Raid Warning Red status when eight Iraqi jets appeared to be heading straight for it. In *London* Commodore Craig immediately contacted General de la Billière at his HQ in Saudi Arabia, informing his boss that he needed permission to fire.

> *When I asked how long I had to take a decision the answer was thirty seconds. So I took it instantly, authorising Chris Craig to defend himself with all the weapons he had.*[9]

Fortunately the Iraqi planes turned away, but within a week squadrons of Allied aircraft would be flying in the opposite direction with deadly intent.

After a further Sea Wolf missile firing, the *London* deployed with the *Argus* to a position near Qatar while the destroyers went further north and *Brazen* headed south to guard the supply ships. The main function of the coalition fleet was to deny Iraqi forces the use of the sea. The enemy would be prevented from making attacks or resupplying their ground troops in Kuwait, the US-UK naval units in the northern Gulf would be able to conduct bombardment missions against Iraqi positions and they could also launch an amphibious assault. The Royal Navy could be expected to use its plastic-hulled Hunt Class Mine Counter-Measure Vessels (MCMVs) close in shore to clear mines both for the amphibious assault and battleship bombardments. The US Navy had very few vessels capable of this very demanding role and none as advanced as the Hunts. Aggressive search and destroy patrols by British Lynx helicopters armed with Sea Skua missiles could be decisive. The ships of the Royal Fleet Auxiliary were responsible for supplying warships of the coalition with everything from toilet rolls to ammunition and would be the only vessels of that kind to risk the dangerous waters of the northern Gulf.

Exocet!

The deadline for Iraq to withdraw from Kuwait or face military action was set for midnight on 16 January. As the sands of time ran out, Commodore Craig sent a signal to the task group ships:

> *We are about to engage in the greatest endeavour of our lives, on behalf of most of the nations on earth...Over the days ahead, total concentration and sustained application will ensure an emphatic victory. May God go with you as you move in harm's way...Without further orders, all Commanding Officers of Royal Navy ships are, on the outbreak of hostilities, to hoist battle ensigns.*[10]

At 2.30 a.m. Baghdad time on 17 January, American ships unleashed the first Tomahawk cruise missiles, which sped to their targets in Iraq, followed by waves of strike aircraft. Commodore Craig saw the spectacle unfold on the radar displays in *London*'s Operations Room:

> *I am watching the greatest single-wave assault of destructive power since the Second World War. A mass of aircraft inch their way implacably across the display towards the top left-hand*

How the Royal Navy's task group was dispersed as hostilities commenced. *C. Craig Collection.*

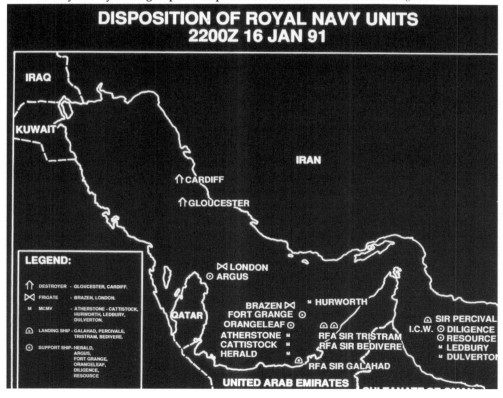

corner – Kuwait and Iraq. Cruise-missile tracks mesh with carrier-launched and shore-based bombers. In sophistication and scale, it is simply unprecedented.[11]

In the *London*'s Sound Room the Commodore's staff were updating their situation boards and keeping tabs on the radar and communications traffic. Chief Petty Officer Cunningham was on watch with his team when Desert Storm was unleashed:

That first night was amazing. The Ops Room displays had hundreds of contacts, so many in fact that they ran out of track numbers.

Petty Officer Burton poked his head in the Sound Room to see if he could help out:

I was fast asleep...the alarm went off and I was instantly awake. By this stage we were sleeping in our clothes, so I leapt out of my bunk and went immediately to see what was happening, but I was sent back to bed.

In the first twenty-four hours of Operation Desert Storm, the Allies flew 1,000 bombing sorties and unleashed 151 cruise missiles, beginning the process of destroying Iraqi combat units in Kuwait and subjecting Iraq's infrastructure to a precise process of dismemberment. In *London* Bob Burton found that he was a man worth knowing...if only he had the authority to tell people what was going on:

Those of us seconded to the Flag Staff certainly knew more than the average sailor. You might go down to the mess for a meal and pick up on a vibe that people were feeling left out of the loop on something. You couldn't tell them yourself, but you could go to the officers and suggest that they should give out some info.

The Iranian Radar Shadow

The American Commander-in-Chief of the multi-national naval task force asked the British commander to put his Type 42 destroyers near the coast of Iran. Commodore Craig, who had seen action as a Type 21 frigate captain in the Falklands' War, was immediately worried about a repeat of the loss in that conflict of the Type 42 destroyer HMS *Coventry*. Close in to the Falkland Islands, she had found herself helpless against Argentinian Skyhawk bombers, because background clutter neutralized her air defence radar. Manoeuvring to solve the problem had left her even more exposed, for *Coventry* crossed paths with *Brilliant*, leading to the frigate's Sea Wolf system shutting down (a problem that had been ironed out by 1991). In the Gulf, Iraqi jets deciding to attack from the direction of Iran would be able to use the background clutter of the coast to obscure themselves, possibly preventing the Type 42s' long-range Sea Darts from getting a good lock-on. By the time the enemy aircraft were within the range of the Sea Wolfs carried by *London,* it might be too late to avoid losing a ship. Also, because the coalition wanted to play by the rules, to avoid Iran being drawn into the conflict, the British ships would not be able to fire while enemy planes were still in Iranian airspace. Luckily the Type 42s were soon sent away from the Iranian coast, to loiter off Kuwait with three US Navy warships as the main anti-air picket line for the task force.

Commodore Craig's new worry was that the Iraqis would try and use the cover of Allied aircraft returning from raids on Baghdad to mask themselves. This duly happened on 24 January but, luckily, an American airborne early-warning aircraft detected the four Iraqi aircraft very early in their attack run. Next the American Ticonderoga Class missile cruiser USS *Bunker Hill*, together with HMS *London* and the Type 42s, picked them up on search radars. But, just as the two British destroyers were about to push the fire button on their Sea Darts, and *London*'s Sea Wolf missile launchers were training towards the danger area, American and Saudi fighters shot down the Iraqi jets. At the height of the action, Commodore Craig had found himself smiling at the cool efficiency with which the Iraqi aircraft had been dealt. He was impressed by the comprehensive layered defence that had been so absent in the Falklands nine years earlier:

> The Iraqis sent two Exocet-carrying Mirage F1s at low-level that were working with Mig 23s as a higher-level distraction. American aircraft took out the high level threat while the Saudis dealt with the F1s. It was very neatly done. There were so many layers of defence. If the Iraqi aircraft directly threatening us had got through the Saudi F15s, then the US Navy's F14s would have been their next hurdle. Only then would we have been expected to deal with them. Everybody was on Actions Stations, all missiles ready. I didn't really feel anxious about it.

The Americans feared that this was a probe and a mass attack would come the following day. When Iraqi aircraft started flying to Iran their fears increased. Were the Iraqis about to try and use the radar shadow of Iran to make their big foray?

In *London* Chief Petty Officer Cunningham wondered if everyone's worst fears were about to come true:

> After the balloon went up on 17 January you thought: 'Well this is it...any moment now, the Iraqi Air Force is going to come straight over the horizon and all hell will break loose, like the Falklands, only much worse'.

But, nothing happened. The Iraqis were seeking refuge for their surviving air force and the fugitive strike jets stayed on the ground for the rest of the war.

The Battle of Bubiyan

On 29 January, as Iraqi forces seized the Saudi border town of Khafji in their only offensive ground action against the coalition, HMS *Gloucester*'s Lynx was patrolling off the coast and picked up a gaggle of radar contacts. These turned out to be seventeen small craft carrying Iraqi Special Forces troops, filtering out of the Bubiyan Channel at the head of the Gulf and intent on a raid from the sea in support of their comrades in Khafji. The Royal Navy's response was devastating. Helicopters from *Cardiff* and *Brazen* joined *Gloucester*'s, going into action by leapfrogging forward using Royal Navy destroyer and frigate flight decks as

Killer and victim – a Royal Navy Lynx flies by a destroyed Iraqi patrol vessel. *Toby Elliott Collection.*

HMS *London* carrying out a RAS with an RFA tanker during the Gulf War. *Toby Elliott Collection.*

staging posts. It was a clear day, with US naval helicopters locating targets while the British did the killing.

> *...Sea Skuas cut low furrows across the glittering waters...*[12]

Carrier-based American A6 Intruder bombers, US Marine Corps Cobra attack helicopters and RAF Jaguars also pitched in.

Over the next thirty-six hours more Iraqi vessels came out, either to go down the coast to Khafji or make an escape bid towards Iran. They were destroyed, with British Lynx helicopters again to the fore. Commodore Craig ruled out an Iraqi attack on *London* and the other Allied warships:

> *If they intended a co-ordinated attack, it is unlikely that they would have chosen to transit over 100 miles south to hit us in broad daylight.*[13]

It was a wild, confused fight, the enemy vessels twisting and turning among oilrigs and suddenly darting out of the haze, or playing dead in the hope of avoiding destruction. Most were left as burning hulks.

During the Battle of Bubiyan, as it became known, twenty-five Sea Skuas were fired and eighteen hit home, with at least seven Iraqi vessels sunk. Several others were badly damaged and finished off by roving Allied strike jets. The Royal Navy's helicopters wiped out a quarter of the Iraqi Navy in those two encounters. The main problem the British helicopters faced was having to stay within enemy radar range while the Sea Skuas rode a beam to their target. This made them very vulnerable to enemy fire, but, as the Iraqi boats increased speed, vibration tended to throw their fire control radar off, so the British helicopters escaped.

British warship numbers rose in late January, with the Type 42 destroyers *Exeter* and *Manchester* together with the Type 22 frigates *Brave* and *Brilliant* arriving in theatre. The mine-hunter force was also boosted with two more Hunts and the survey vessel *Hecla*.

The commanding officer of HMS *Brilliant*, Captain Toby Elliott, had made sure that his sailors were kept sharp from the moment they left the UK:

> *At the time I thought that it was a long way to the Gulf and it might all be over by the time we got there. In any event, we were determined to make sure that we would be ready for*

whatever the Iraqis might throw at us. The most serious threat was from their air force and we had plenty of air defence exercises on the way to the Gulf. In the eastern Med we passed through the Ark Royal group where I think there was a hint of frustration. They must have sent us on our way with mixed feelings. The thing that focussed our minds more than anything else as we got closer to the Gulf was the Chemical and Biological threat. After entering the Straits of Hormuz we started to notice the huge coalition naval force. There was such a large presence; in fact there were always one or two other naval vessels in sight. Our main job was to stay with the supply vessels of the fleet train in the central Gulf. This meant staying off Bahrain, but from time-to-time we did escort the fleet train up north to replenish the other warships.

Mines Claim Two Warships

As January gave way to early February, more small craft were detected making runs down the coast and were destroyed by helicopters from *Manchester* and *Cardiff*, while *Gloucester*'s Lynx came under fire from Iraqi anti-aircraft guns ashore. But, after the Battle of Bubiyan, the only threat to the task force was from mines and a few missile batteries on the Kuwaiti coastline.

The main task became clearing 'boxes' of sea to enable American battleships to get close enough to carry out accurate bombardments.

By early February sightings of floating mines were becoming more frequent, but the Allies were fairly sure they knew the limits of the Iraqi minefields off Kuwait. The naval command felt the lead elements of the task force could go forward and begin clearing paths into them with some confidence.

On 14 February *London* led the British task group forward in company with the American assault carrier USS *Tripoli*, plus the *Wisconsin* and *Missouri*, the air-defence cruiser USS *Princeton* and some old wooden-hulled American minesweepers. Almost the entire British naval force was integrated with the Americans for this hazardous phase – of thirty-four

Iraqi mines like this one, pictured drifting in the northern Gulf, damaged two American warships and could easily have sunk HMS *London*. *C. Craig Collection.*

vessels half were British, but the whole group was under the tactical command of the Americans, with the *Tripoli* acting as flagship. *Brazen* and *Cardiff*, long overdue for relief, had by then sailed for home, but Commodore Craig felt he could not allow his flagship or *Gloucester* to leave at that point. *London* and *Gloucester* were fully worked up, providing close cover for the Hunt Class minehunters and HMS *Herald*. On 17 February active Silkworm batteries were detected trying to lock onto the US-UK naval force, so the Americans ordered a withdrawal, over which Commodore Craig had some misgivings:

> *I felt we might have stood our ground. The Silkworm is big and slow, an ideal target for the Sea Dart. If we withdrew at every alarm we would not reach the coast before Christmas.*[14]

The task force moved back in towards the coast when it looked as if the threat had receded, but just before 2 a.m. on 18 February the USS *Tripoli* hit a mine.

> *The reverberations thudded through* London *three miles away, leaving no possible doubt as to what had occurred. Somebody had 'bought it'.*[15]

The *Tripoli* had fallen victim to a tethered mine, which meant the Iraqi minefields extended further than expected. The ships stayed where they were, until mine warfare vessels could lead them out of danger. *Tripoli* stayed afloat, as she was a robust ship but Commodore Craig was all too aware that a mine strike might well sink *London*.

> *...I reflected upon the loss of life if a similar explosion were to rip through my flagship, whose hull was but a fraction of* Tripoli's *half-inch steel skin.*[16]

To fix the minefield, they needed to find another mine and while HMS *Brave* had spotted mines drifting perilously close to her hull, they were untethered floaters. Two-and-a-half hours after the first explosion, the USS *Princeton* detonated a mine.

> *At 0420 another massive reverberation echoed through* London *and again all ships slid to a halt...*[17]

Petty Officer Burton was in the Sound Room exchanging information with Fleet HQ back in the UK:

> *The whole ship shook and I typed in: 'I have to go now'.*

Chief Petty Officer Cunningham found his sweet dreams rudely interrupted:

> *I had just gone off-watch and was in my scratcher when the mine went off. It lifted the back end of the ship about six inches ...it was a hell of big bang. I heard it and felt it. The Chief in the bunk below me said: 'What the F*** was that!?' We got to our Action Stations as quickly as possible.*

In both mine strikes there were surprisingly few casualties and no deaths, but the *Princeton* was so badly damaged she had to be towed to a dockyard in the southern Gulf while *Tripoli* withdrew for repairs under her own steam a few days later.[18]

Behind them, the two American warships left a new sense of vulnerability. For Chief Petty Officer Cunningham in the British flagship, the reality of the situation hit home.

> *You were so busy you didn't pause to think about anything outside your own little world, but something like the* Princeton *getting hit really brought it home to you – this was a real war and people could get hurt.*

The Iraqis Strike Back

On 22 February, realizing they would not be able to hang on to Kuwait, the Iraqis began torching the sheikhdom's oil wells, with 150 out of 950 ablaze by the following morning.

To the south of *London*, a couple of Silkworms were fired at *Wisconsin* but they missed, even though the battleship made the biggest of maritime targets. Her sister ship *Missouri* was keeping close company with *London*, *Gloucester*, the Mine Counter-Measures Vessels

The British naval task group, with *London* (far right). *Toby Elliott Collection.*

and the guided-missile frigate USS *Jarret*, which had replaced the *Princeton*. Finally the naval forces were hitting the heart of the occupiers, as the *Missouri* '...unleashed the immense destructive power of her guns upon Faylakah Island...The shockwave ran through all the ships psychologically and physically.'[19] Faylakah guarded the seaward approaches to Kuwait City and on 23 February the Americans dropped a pair of 12,000 lb bombs on its unfortunate defenders.

On 24 February, following probes along the Saudi-Kuwait border, the coalition unleashed its massive ground assault. Seven hundred thousand Allied soldiers and 3,400 tanks were

An Iraqi Silkworm missile. *Toby Elliott Collection.*

pitted against 180,000 Iraqi troops and around 2,000 tanks in Kuwait itself, with more enemy units lurking just inside Iraq.

By the early hours of 25 February, as the Iraqi Army crumbled and Kuwait City was liberated by Arab troops and the US Marines, the *Missouri* and *Wisconsin* were firing on enemy positions on Faylakah and against armoured formations and supply dumps around Kuwait International Airport.

The *London*'s sailors marvelled at the terrible majesty of the battlewagons letting rip with their big guns. Even Chief Petty Officer Cunningham managed to escape from the Sound Room to do a little sightseeing.

> *The first couple of weeks we were too busy for me to get out on the upper deck to see what was happening, but, as the war went on there were less aircraft flying around and the Iraqi threat at sea diminished, so I got more opportunities. It was incredible watching the flashes ripping the night air apart and feeling the shock wave wash over me.*

At 1.51 a.m. on 25 February, the Iraqis hit back, firing two Silkworms at the naval task force. The big lumbering beasts followed a flat trajectory out to sea but one went haywire and splashed shortly after launch. The other '...continued at a little less than the speed of sound and at 1,000 feet straight towards *Missouri*'.[20]

The contact that appeared on the radar displays in *London*'s Ops Room could easily have been one of the many Allied aircraft that frequently transited through the combat area unannounced. But *Gloucester*'s Ops Room decided this was something different:

> *She alone of the air-defence ships detected the missile, correctly identified it, allocated it to Sea Dart, ensured that no aircraft were in the vicinity, and obtained command approval to engage.*[21]

Two Sea Darts leapt off the destroyer's launcher rails and sped away at Mach 3. Everyone in *London*'s Ops Room, and looking on from the frigate's bridge and upper decks, held their breath:

> *It was a perfect shot, and the big Silkworm erupted amidst a bright blue flash*[22]

Only three minutes had passed between the Silkworm being launched and *Gloucester*'s Sea Dart destroying it. In the frigate's Ops Room Commodore Craig heard that some of *London*'s watchkeepers feared she was under multiple attack:

> *The wreckage of the Silkworm dropped very close to* London*'s stern. People on watch on the upper deck saw the splashes and thought we were under fire from rockets, but it was the various bits of wreckage.* London *had been under the missile's flight path.* Gloucester *was up-threat, fired her missile...obtained a hit...and the debris continued on the line of flight. After the war I learned that the Iraqis were targeting the last known position of the battleship.*

According to Yorkie Cunningham *Gloucester* wasn't the first to fire at the incoming Silkworm.

> *An Able Seaman on HMS* Herald *was on the upper deck and saw this light coming, realised it was an enemy missile, cocked his 20mm gun and fired at it. The Hunts were the closest to the coast and the* Herald *was next. Our Sea Wolf never even got a sniff. The thing they reckon saved us was that the Silkworm's internal search radar was not on, allowing* Gloucester *to get it before it got us.*

It was the first combat interception of a missile by a warship-launched missile in the history of naval warfare. In firing the missiles, the Iraqi battery gave away its position and American naval aircraft swooped in to destroy it.

A Vision of Hell

The main value of the huge naval force arrayed off the Kuwaiti coast was in tying down thousands of Iraqi troops, who believed that a powerful amphibious assault was going to be launched against them. However, it was never a serious proposition, so the Iraqis sat and waited in vain while the Allied land offensive got underway and drove them from the raped sheikhdom.

Sailors in HMS *London* and the other task force ships were aghast at the sheer wanton vandalism of the Iraqis, who detonated a further 350 wellheads as they fled. The vista that stretched before the coalition warships was a scene from hell.

> *By night, the blanket of smoke would lift just enough to show us the carnival death-dance of the oil fires ashore. From horizon to horizon, the entire coast seemed ablaze, with only the silhouettes of the Hunts etched defiantly against the flickering amber glow. On this Satanic stage, the elevated gun barrels of the battleships traversed back and forth, seeking fresh prey, before erupting their own torrent of flame – as if hungry to join the inferno. We were at the edge of hell...[23]*

One newspaper correspondent mournfully noted:

> *Apocalypse now. A sickly, mournful shroud of smog permanently masks the sun, plunging war torn Kuwait into perpetual twilight...Black rain pours down from the sky and the once calm, beautifully clear, blue waters of the Arabian Gulf are ruined by the lifeblood of the world's economy...[24]*

Job done, HMS *London* handed over flagship duties to HMS *Brave* and headed home to the UK on 28 February, the day President Bush called a halt to the Allied ground offensive. Commodore Craig stayed with *Brave*, sad to see his former flagship go:

> *I bade farewell to the splendidly loyal* London *and watched her turn south at last.[25]*

For their part the Sound Room team in the frigate were sad to go back to their day jobs as they had enjoyed working for SNOME:

> *Commodore Craig was a really good guy, just the sort of charismatic leader you always imagine the Royal Navy will produce in time of war. He got the nickname 'Captain Charisma'*

An American battleship unleashes hell against Iraqi troops in Kuwait. *Toby Elliott Collection.*

and he was too nice to treat it seriously. He didn't want that title, but I am afraid that he earned it, so it stuck, much to his embarrassment.[26]

As *London* sailed home Commodore Craig went ashore to discover the scale of devastation for himself:

Driving through Kuwait City within hours of the cessation of hostilities was like entering a little bit of Hell. But the Ahmadei oil fields were undoubtedly the worst. The palls of smoke overhead and gloom during the day created a place darker than the darkest night. Kuwait's port areas had all been levelled, bodies were floating in the waters. Oil slime covered everything. I didn't see any large-scale casualties like they did on the Basra road, but it was still a very vivid and poignant experience. It was a very depressing spectacle. When you are at sea watching the war develop on radar...you don't really see the full price paid.

Aside from the battlefield carnage a few hundred Allied troops had been killed, but many thousands of Iraqis lost their lives including around 30,000 during the air phase of the campaign alone.

The Legacy of War
The Royal Navy's reputation as one of the world's pre-eminent fighting forces had been reinforced by the conflict in the Gulf.

American liaison officers on British ships were impressed by the degree of integration of the electronic systems aboard. The US Navy, being much larger, still had highly specialized, less versatile ships, whereas the Royal Navy had been obliged to opt for almost interchangeable destroyers and frigates. The Americans were also surprised, on offering the British access to data from their satellites to find out that they had already deciphered the signals...[27]

In fact, Royal Navy sensors were so effective they were routinely able to pick up the so-called 'invisible' stealth bombers of the US Air Force as they headed for Baghdad.[28]

For Chris Craig the conflict in the South Atlantic of 1982 had been an invaluable baptism of fire for his Gulf War task group:

Combat experience is like no other conditioner. Sixty per cent of my COs had been warfare officers in the Falklands War. We knew the business and we had an underlying confidence other nations did not have. Not even the Americans could rival our experience at close in-shore fighting. We were leaner, meaner and fitter, with damage control and chemical warfare defences second to none. Gloucester's missile kill, the MCMVs' clearance rate, the performance of the Lynx helicopters armed with the Sea Skua missile, all confirmed that diagnosis. Our sailors themselves were magic. The youngsters had to go through their own conditioning, but there was an underlying belief in what we were doing and the war experience of their senior officers and senior rates created confidence. They thought 'my Captain has been there, my Chief has done it too, so I should be able to handle it'... I think there was a huge worry over chemicals, the media made a great deal of it, but it was vastly over-stated, an undue anxiety. I think it was a Just War and we can be proud of what we did. We drove the Iraqis out of Kuwait and righted a wrong.

What the Allied ground attack didn't do was destroy the Republican Guard. It continued to maintain Saddam Hussein's regime, which was '...cruel, lying, intimidating, and determined to retain weapons of mass destruction...capable of killing thousands, even millions, at a single blow...'[29]

Notes

1. General H. Norman Schwarzkopf, *It Doesn't Take a Hero.*
2. Ben Brown and David Shukman, *All Necessary Means.*
3. Norman Friedman, *Desert Victory.*
4. Captain Chris Craig, *Call for Fire.*
5. *Evening Herald*, Plymouth, 31 December 1990.
6. The author writing for the *Evening Herald,* Plymouth, 8 January 1991.
7. ibid.
8. ibid.
9. General Sir Peter de la Billière, *Storm Command.*
10. Captain Chris Craig, op.cit.
11. ibid.
12. ibid.
13. ibid.
14. ibid.
15. ibid.
16. ibid.
17. ibid.
18. ibid.
19. ibid.
20. ibid.
21. ibid.
22. ibid.
23. ibid.
24. The author in a feature for the *Evening Herald,* Plymouth, 20 March 1991.
25. Captain Chris Craig, op.cit.
26. Yorkie Cunningham in interview with the author, November 2001.
27. Dan van der Vat, *Standard of Power.*
28. Antony Preston, writing in *WARSHIPS IFR* magazine, Winter '99 edition.
29. Richard Butler, *Saddam Defiant.*

Chapter Twenty

THE FALL OF THE SOVIET UNION

Home are the Heroes

The *London* sailed into Plymouth Sound on 25 March, making a triumphal progress past the Hoe. She was escorted by tugs spraying fire hoses and RAF Jaguar jets swooped low overhead to salute her return from war. The overjoyed friends and relatives of her sailors packed the jetty at 'Frigate Alley' in Devonport Naval Base.

> *Descending into the carnival atmosphere on the quayside, the men were instantly swamped by tearful relatives and wellwishers.[1]*

While her sailors enjoyed post-conflict leave with their families, the frigate entered an Assisted Maintenance Period (AMP).

Following local running in May and June, to bring her systems and sailors back up to scratch, it was indicated to HMS *London* that she should start preparing to be Britain's ambassador on a special mission. The fiftieth anniversary of the passage of the Operation Dervish convoy to north Russia was approaching and, because of the name's past associations, *London* was thought an ideal vessel to represent the Royal Navy.

The commemoration had been a year in the making, with war veterans from the UK's North Russia and Russian Convoys clubs working alongside their Soviet counterparts, and the Russian government, to arrange a series of events. An Operation Dervish '91 'convoy' to Murmansk and Archangel was to be the centrepiece. To show its significance, UK Prime Minister John Major and Soviet President Mikhail Gorbachev were to be joint honorary Operation Dervish commodores.

An Operation Dervish '91 badge distributed by the Soviets. *L. Abernethy Collection.*

HMS *London* returning to Devonport. *H.M. Steele.*

It was a remarkable plan, for the two northern Russian ports had not been visited by Western naval vessels since the Second World War. In the intervening five decades Murmansk and Archangel had become among the most secret places in the Soviet Union, with the main striking power of the massive Red Navy lurking in bases across the Kola Peninsula and around the White Sea.

The historic Royal Navy visit to the lair of the Russian Bear was all part of Gorbachev's policy of thawing relations between the USSR and the West. It was only after he came to power in 1985 that Russia had officially acknowledged to its own people that the Arctic convoys had taken place. The convoy re-enactment of 1991 would, hopefully, bring to an end the Cold War confrontation between the British and Soviet fleets.

The Hardliners' Coup

Diehard communists within Gorbachev's own government did not welcome the friendly attitude to the West or liberalization at home. The withdrawal of Soviet troops from Afghanistan in early 1989, the beginning of a pull-out from Eastern Europe, plus the overwhelming defeat of Soviet weapons in the Gulf War, had all heaped humiliation on Russia.

In June 1991 Boris Yeltsin had become President of the Russian Republic and on 20 August he was due to sign a treaty with Gorbachev that would leave the Kremlin in charge of foreign affairs and defence while the USSR's republics governed themselves. It was the final straw for the hardliners who decided to take strong action. Gorbachev's decision to leave Moscow for a holiday prior to the treaty signing offered the perfect opportunity to isolate him and seize the reins of power. Putting Gorbachev under house arrest at his Black Sea holiday retreat on 18 August, the coup leaders were reassured by tacit support from Defence Minister Marshal Dmitri Yazov. By 7 a.m. on 20 August tanks and armoured personnel carriers from the Red Army had entered central Moscow. A state of emergency was declared, with the Soviet population being informed that Gorbachev was 'ill'.

As these dramatic events unfolded in Moscow, HMS *London*'s new Commanding Officer, Captain Mark Stanhope, was being advised to carry on with preparations for departing Devonport. The frigate would head north for if the hardliners consolidated their hold on power, a tight watch might need to be kept on the Northern Fleet. The Russian Sovremenny Class destroyer *Okrylenny*, which was due to take part in Plymouth Navy Days, had already turned around and was heading back to base.

Yeltsin, whom the plotters had foolishly failed to arrest, led resistance from the Russian Parliament building in Moscow. Using foreign and domestic television and newspapers, which the incompetent reactionaries had failed to muzzle, he broadcast appeals for people to join him. Having enjoyed the freedom provided under the limited economic and social reforms, the majority of the Soviet peoples willingly responded to Yeltsin's call.

By the evening of 20 August the hardliners were planning an assault on the Russian Parliament. But the young soldiers in the tanks surrounding the building would not fire, some even going over to Yeltsin's side, and even the KGB's crack Alpha Group commando unit refused to storm the Russian Parliament. Marshal Yazov was also being pressurized by the heads of the Navy and the Rocket forces to withdraw support for the coup. With what little support they had fast slipping away, and with thousands flocking to the Russian

Parliament's defence, the hardliners decided they had better go back to Gorbachev and see what could be salvaged from the mess. The Soviet president was released from house arrest while coup leaders were arrested. When he returned to Moscow, Gorbachev tried to pretend that nothing had happened and resume control, but the events of 19-21 August had hastened the Soviet Union's demise and ensured his eventual removal from office.

Voyage Into Uncertainty

On 21 August, HMS *London* had sailed from Devonport and, when it became clear the coup was collapsing and Gorbachev was back in the Kremlin, the UK government decided to send her to Russia as planned. It would be an important signal of support from the West at a decisive moment. Calling in at Rosyth Naval Base, the *London* took onboard a Royal Marine band, selected members of the media and additional naval personnel, including translators and communications specialists.

London sailed out under the Forth Bridge on 23 August, the same day Gorbachev addressed the Russian Parliament and revealed the full scale of the treachery within his own administration. This served to tear away his last remaining shreds of authority and emboldened Yeltsin to push through a decree removing the Communist Party as Russia's sole ruling group. On 24 August the Communist Party was formally dissolved after Gorbachev resigned as its Secretary General. He retained presidency of a rapidly fragmenting Soviet Union. The Baltic States had declared independence on 20 August rapidly followed by the Ukraine, Uzbekistan and other republics.

As HMS *London* sailed north, a great deal of uncertainty still hung over the mission for the Soviet Navy might be paralysed by political upheaval. A Soviet intelligence-gathering trawler, sprouting a profusion of aerials, was at least at sea, trailing the *London* as she progressed around the top of Norway. The Russians also traditionally kept an attack submarine on barrier patrol duty off northern Norway and *London* was ordered not to try and avoid being found. It was also thought likely the Russians would send aircraft over to keep watch and *London* would deliberately not lock on to them with her radar, as had been the practice during the Cold War.

Preparations aboard ship carried on regardless, with sailors and marines '...honing the ship's Ceremonial Entry Procedures for Murmansk and Archangel, the firing of 21-gun salutes, Guard and Colour Party drills, and the playing of the Soviet Union's national anthem'.[2]

On 26 August *London* was joined by the Royal Fleet Auxiliary oil tanker *Tidespring*, which would be her partner for the trip, providing fuel and also representing the UK's Merchant Navy in the convoy. The British naval vessels paused off Tromsø, with *London*'s Lynx flying in to pick up Rear Admiral Bruce Richardson who would lead the UK participation in Operation Dervish '91.

Shortly before 6 p.m. on 26 August, *London* received official confirmation that she and *Tidespring* would be permitted to enter Russian territorial waters for 'bilateral exercises' with Soviet naval vessels. Captain Stanhope told his sailors over the tannoy:

> *Get learning your Russian – and make sure you can at least order a drink for our hosts. I am sure we'll all have an interesting, historic and enjoyable visit to Russia.*[3]

The British sailors also received a printed guide on how to conduct themselves in the alien environment of the Soviet Union:

The guide contained several useful phrases in Russian... in addition to asking for a vodka, HMS London's *sailors were capable of paying their female hosts a compliment, by saying 'you are a pretty girl!' The guide advised the sailors to be patient in their dealings with local people: 'Even simple matters such as entering a restaurant or buying something involves a lengthy process of questions and negotiations.' The sailors were advised not to drop litter in the streets, as it is regarded as decadent behaviour...The guide advised: 'On no account flaunt your wealth while ashore.' Soviet sailors get only a fraction of what RN sailors are paid. For example, an Able Seaman in the Soviet Navy receives five roubles a month – or 12p at current tourist exchange rates.*[4]

The Cold War Façade Crumbles

A Soviet warship was due to rendezvous with the *London* around 9 a.m. on 27 August but she failed to show. Two hours later, as the British naval vessels steamed slowly east, the Soviet Navy's emissary finally turned up.

The spy ship that trailed HMS London *all the way up the Norwegian coast pulled back as the Devonport frigate rounded the North Cape of Norway and entered the Barents Sea. A sinister, fast-moving silhouette on the horizon soon provided the answer. The crew of HMS* London *prepared for their first encounter with the pride of the Russian Navy – the high-tech Sovremenny Class missile destroyer* Rastoropny... *During many years of cat-and-mouse with their Cold War foe, Royal Navy sailors were accustomed to cool receptions, but the stony façade of Kremlin communism was shattered as the* Rastoropny *steamed past – for the Russian warship's crew crowded the upper deck, giving* London *a hearty cheer and waving their caps wildly.*[5]

One of the interpreters standing by on the bridge translated a request from the *Rastoropny* for *London* to proceed to a rendezvous point where she would meet the Krivak Class frigate

HMS *London*'s Lynx pilot watches the *Rastoropny*. *Iain Ballantyne.*

Gromky, carrying Vice Admiral Igor Kasatonov, Deputy Commander of the Northern Fleet. The *Gromky* would lead the British naval vessels to the Operation Dervish convoy.

During the next seven hours Soviet maritime patrol aircraft, including Badger bombers, appeared without warning to make slow passes overhead but, as night drew in, the *London*'s sailors began to wonder if the Russian fleet was still so disorganized *Gromky* and the convoy had been unable to sail.

As HMS London *sailed out of a brilliant orange sunset in the west, the Soviet Navy frigate* Gromky *suddenly plunged out of the steel grey curtain of drizzle and gloom that had fallen in the East. She closed in spectacular fashion – at her top speed of 30 knots, prow plunging into the sea, ploughing a furious furrow of spray. The crew of HMS* London, *who were told to clear lower decks, gave her a rousing welcome, cheering 'hurray! hurray!*

Captain Stanhope (left) and Admiral Richardson greet the *Gromky*. *Iain Ballantyne.*

hurray!' and lifting their caps in salute. The Gromky's *crew replied in similar fashion and with great gusto. 'Normally they're not so sociable... but that's pretty impressive,' remarked Petty Officer Chris Smith. Then all eyes switched skywards as two Soviet naval air arm SU-27 Flanker fighters wheeled low overhead with a bunsen burner roar. They dipped their wings in salute before their pilots pulled back hard on their control columns, showing off what fantastic technology the Soviets could produce to the old enemy... taking their jets up in a spectacular climb above the British frigate. Soon HMS* London *was cloaked in black, red and grey smoke, created by Soviet naval vessels only faintly made out in the gathering night. Flares dropped*

An aerial view of the *Gromky*. *L. Abernethy Collection.*

from low flying Mail anti-submarine flying boats and May maritime patrol aircraft created more smoke. When the smoke screens dispersed London *found herself passing the Soviet Navy hospital ship* Svir, *which was carrying more than 100 British and Russian veterans of the Arctic convoys. The cheering veterans welcomed HMS* London *to the fold, shouting and waving from the guardrails of the* Svir's *packed upper decks.*[6]

One of the British veterans, who had flown out to Murmansk to board the *Svir*, later wrote of this moment:

As darkness approached once again the red sky appeared to the north and then we saw the shape of HMS London *closing with the RFA* Tidespring. *This brought a lump in the throat and a great deal of pride, to think that here was one of our own warships coming all this way, as we liked to think, to honour some battle-hardened old veterans who weren't ashamed to shed a tear.*[7]

In addition to naval ships there were half a dozen Soviet merchant vessels in the convoy and also gathered around *London* were the *Gremyashchy* (sister to the *Rastoropny*) and the Kashin Class destroyer *Smyshlenny*. After decades of jostling for the freedom of the seas as Cold War foes it was all a bit too surreal for some of *London*'s sailors, including Yorkie Cunningham:

As I cheered them earlier, I definitely felt mixed emotions. These people had technically been our enemy since the late 1940s and then all of a sudden we were expected to greet them with open arms. As a sonar rating I had trained to find and identify Russian warships for possible destruction if things turned hot. But here they were that evening at very close quarters – our new pals.

Soon there would be an incident that, for a moment, made the *London*'s sailors think their new found mates were not so friendly after all.

'Torpedo!'

The following morning promised to be action-packed, with the fiftieth anniversary convoy subjected to the re-enactment of a German attack.

In order to start things off with a bang, a Tango Class submarine surfaced close to *London* and, to the amazement of those looking on from the British warship, two hapless Russian sailors erected a saluting gun. Despite being washed off the casing a couple of times, with one sailor hauling the other back up by his safety line, they managed to fire off a few blank rounds. Dismantling the saluting gun, the sailors retreated inside the submarine, which promptly dived. In *London*'s Operations Room sonar operators kept track of the Tango's movements. Aircraft swooped low, dropping smoke flares to simulate German bombing runs and Russian warships fired back with blanks in self-defence. *London*'s Lynx helicopter took some members of the embarked media group across to the *Svir*, making history as the first British naval helicopter to land on a Soviet Navy vessel. A little while later a correspondent, chatting to war veterans aboard the *Svir*, found himself distracted:

A Soviet flying boat flies over HMS *London*.
Iain Ballantyne.

A Russian Navy Helix helicopter circles the *London* as she makes her way down the Kola Inlet. *Iain Ballantyne.*

> *Torpedo tracks to starboard stopped further conversation with one heading astern of the* Svir *from which the torpedo suddenly leapt out of the water. HMS* London *took avoiding action and the RFA* Tidespring *laconically reported later that she thought she was about to be the first ship to be torpedoed in the Barents Sea since the Second World War!*[8]

Another correspondent, who had preferred to stay in HMS *London*, reported it from the viewpoint of the frigate's crew:

> *They were convinced the torpedo would find its mark but in the nick of time* London *took avoiding action, and it passed by 40ft from her bows...One sailor aboard* London *said: 'We saw this red thing bobbing to the surface and then realised it was some kind of self-propelled practice torpedo as it started to head for us.'*[9]

London's sonar operators had assumed the submarine would just shadow their ship to replicate a U-boat attack:

> *But they got the shock of their lives when the sub fired the torpedo – and it came straight for them. The Royal Navy's practice torpedoes have no propulsion and merely float to the surface after a practice firing, but the Soviet missile was wire guided and clearly powered.*[10]

One of *London*'s officers recalled that prior notice was given, but it failed to prepare the frigate for the reality of 'practice' Russian-style.

Sailors on *Gromky* were friendly, if cautious, hosts when *London*'s sailors paid them a visit while both warships were alongside in Murmansk. *Iain Ballantyne.*

I think the interpreter was told by the Gromky *that the Russians were going to fire some dummy torpedoes at us. We weren't totally clear what that might involve until we heard the* Tidespring *on the radio saying: 'Someone's trying to torpedo us!'* [11]

The Russians themselves were pretty astonished by the incident.

A Soviet Navy officer later admitted to me that the rogue torpedo had gone out of control through a failure of its guidance and depth controls: he had never seen anything like it before. [12]

Russian warships, with their welded steel hulls can easily shrug a propelled practice torpedo off. However, the tensile steel hulls of modern British frigates are so thin a practice torpedo could easily have penetrated the *London* and, possibly, sunk her.

Into the Lair of the Bear

Now came the moment of maximum intrigue – the passage of a British warship,

30 miles up the Zaliv River to the fishing port of Murmansk. The crew of the Type 22 frigate passed the most secret bases in the Soviet Union – the first Western warship ever to get a close look at naval installations that simply don't exist on maps. [13]

For Sub Lieutenant Lee Abernethy it was a fascinating vista:

The Kola Peninsula appeared very desolate and bleak. It was very mysterious. We were very impressed by all the ships in there and the extensive dockyard facilities. We thought: 'If they have this many in refit they really must have a lot of warships.' Of course we didn't know the reality of it all, which was that a fair proportion of their fleet couldn't put to sea.

Yorkie Cunningham was similarly awestruck:

It was very eerie. Alongside in their dockyards were warships with bits of steel cut out of them and a fair few November and Charlie Class submarines tied up. A Kirov Class nuclear-powered battle-cruiser in a dry-dock made a really impressive sight – absolutely massive and looking very deadly. All that naval hardware seemed to go on forever.

Chief Petty Officer Chris Norris found the experience amazing:

To be able to actually go down the Kola Inlet was quite something, especially for someone like me, having spent a fair bit of time in warships shadowing the same vessels.

On approaching the carrier *Gorshkov*, flagship of the Northern Fleet, off Severomorsk, *London* fired a 21-gun salute, while the Royal Marine band struck up the Soviet anthem. The *Gorshkov* replied in kind.

One of *London*'s sailors and a Soviet rating stand side by side in salute to the sacrifice of convoys to Russia in the Second World War. *Iain Ballantyne.*

HMS *London* fires a 21-gun salute as she arrives at Archangel. *S. Reed.*

Going in stern first alongside a jetty at Murmansk, to a soundtrack provided by the band of the Northern Fleet and the Royal Marines, *London* was placed next to the *Svir* and *Gromky*. In addition to a number of civic receptions organized in their honour, *London*'s sailors got to go on a run ashore in their best uniforms – something they were unable to do at home in the UK at that time because of the terrorist threat. Bob Burton was on gangway duty, responsible for checking the sailors returned in good order.

The hospitality was wonderful in Murmansk, which was obvious when members of the ship's crew came back to the ship. If officers and senior rates could salute after they came up the brow then they were OK. Likewise, if the junior rates could say 'Good evening sir...' then they were OK too. It was quite obvious that the ordinary Russians were on a mission to impart their own personal brand of Glasnost to our boys.

Chris Norris found the Russians to be incredibly warm-hearted.

Murmansk was a very grim place but its residents would have given you their last rouble. We bought some people a coffee in a bar and they couldn't believe it. When we met them the following day they gave us fruit, which was a luxury to them.

When *London* left Murmansk on 29 August, in company with the *Gromky* and *Svir*, the latter carrying British and Russian war veterans, she had some extra passengers onboard. Ten British war veterans sailed with the frigate to Archangel, with one of them dropping a wreath over the side during a remembrance service held en route. Entertaining the veterans in the Chief Petty Officers' mess enabled Yorkie Cunningham to compare his more recent experiences of war with theirs:

When they told us about what they went through on the Russian convoys it brought home to us that the Gulf War was pretty minor stuff. There was no comparison.

After sailing through the White Sea, *London* followed the two Russian vessels up the river Dvina to an incredible welcome at Archangel on 31 August, fifty years to the day since the original Operation Dervish convoy arrived.

> *For the many Russians who lined the waterfront, or brought their yachts and motorboats out to dance around the frigate in greeting, she symbolised the end of the regime brought to power by the 1917 revolution. Aboard one of the boats a woman held a megaphone to her lips and shouted in greeting: 'Sailors of HMS London, welcome to Archangel – you have liberated us!'[14]*

Treated like movie stars by a jubilant population, the British sailors were asked to sign copies of a newly published book that detailed the heroic sacrifice of the Arctic convoys. The young sailors of 1990s *London* were receiving gratitude denied to the men who crewed her forebear during the Second World War.

'...you have liberated us!' was the cry from the yachts that danced around HMS *London*. *Iain Ballantyne.*

> *It was simply overwhelming – you couldn't move for overjoyed Russians.[15]*

Captain Stanhope received an unusual request that he was delighted to fulfil – the christening of a baby in *London*'s ships bell. It came from Paul Wyatt, a former Royal Navy submariner who had married an Archangel girl called Natalie and moved to the Soviet Union in 1990, when it looked like Gorbachev's reforms were making life easier. The christening of their eight-month-old son Alexei, took place on *London*'s bridge, a moment that symbolized hope reborn following the black days of the hardliners' coup.

Notes

1. *Western Morning News*, 26 March 1991.
2. Major General Edward Fursdon, in a feature article entitled *Operation Dervish '91* published in *Navy International*, October 1991.
3. The author writing for the *Evening Herald*, Plymouth, 27 August 1991.
4. The author writing for the *Evening Herald*, Plymouth, 30 August 1991.
5. Diary notes made by the author aboard HMS *London*, during the voyage to Murmansk, August 1991.
6. ibid.
7. W. Smith writing in *Northern Light*, the official publication of the North Russia Club, December 1991.
8. Major General Edward Fursdon, op.cit.
9. The author writing for the *Evening Herald*, Plymouth, 29 August 1991.
10. ibid.
11. Lieutenant Commander Lee Abernethy in an interview with the author.
12. Major General Edward Fursdon, op.cit
13. The author writing for the *Evening Herald*, Plymouth, 30 August 1991.
14. The author writing for the *Evening Herald*, Plymouth, 16 September 1991.
15. Bob Burton in an interview with the author.

Chapter Twenty-One

KEEPING WATCH ON THE BALKANS

Saddam Remains Defiant

Arriving back at Devonport on 9 September, the frigate departed on a Western Atlantic (WESTLANT) deployment in company with one of the Royal Navy's Invincible Class carriers little more than a week later. After two months of exercises with the US Navy, the *London* headed for home in December but a diversion created a tricky situation.

> *One of our guys fell ill, so we diverted to Halifax and flew him off. However, this meant the ship had travelled much further north than anticipated. The weather was a bit heavy in northern latitudes and this, combined with the diversion, meant that when we got back to Devonport we had just 13 per cent fuel left. We were concerned that we might have to get an RFA tanker out to give us a top up to get home.*[1]

HMS *London* (left), the oiler RFA *Brambleleaf* (centre) and destroyer HMS *Newcastle* composed the Royal Navy's Gulf task group in early 1993. *T. McClement Collection.*

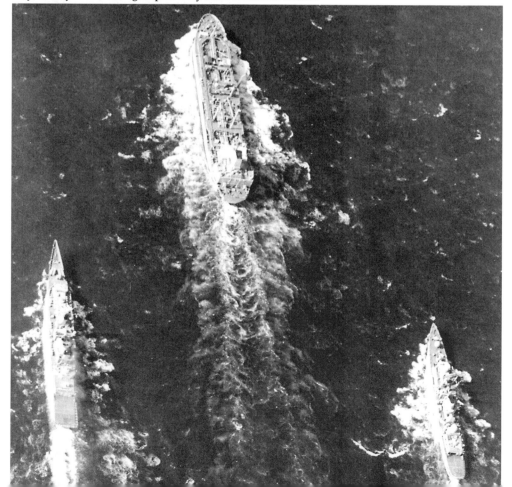

The frigate's boarding teams practise their rapid-roping techniques on her forecastle during the 1993 deployment to the Gulf. One of *London*'s principal missions was enforcing UN trade sanctions against Iraq by stopping and searching suspect merchant ships. *T. McClement Collection.*

The *London* makes a high speed pass in the Gulf. *T. McClement Collection.*

After another period of maintenance the *London* went through Operational Sea Training before embarking on a series of NATO and Royal Navy exercises in the Atlantic. After a visit to the city of London in June 1992, the frigate called at the French naval port of Brest before returning to Plymouth to prepare for a deployment to the familiar waters of the Gulf, where trouble was brewing again.

Having kept his head down for a while, Iraqi dictator Saddam Hussein had ordered his officials to become more aggressive in their evasion of ceasefire conditions laid down by the United Nations following the defeat in Kuwait. Saddam was particularly keen to avoid UN inspection teams finding and destroying his weapons of mass destruction or the infrastructure supporting them. By the time Captain Tim McClement took over command of HMS *London* from Captain Stanhope, at Singapore on 29 December, matters were beginning to reach boiling point.

The frigate was heading the Royal Navy's long-established Armilla Patrol, which also consisted of a Type 42 air-defence destroyer and an RFA oiler. The warships rotated in and out of the Gulf to give their crews a break from the monotony of enforcing the UN sanctions. In January 1993 HMS *London* was enjoying a port visit to Chittagong in Bangladesh when Captain McClement received an urgent signal:

> We were due to go to Colombo, but were told to get back to the Gulf ASAP because something was going to happen.

On 17 January, the second anniversary of the beginning of Operation Desert Storm, American naval forces unleashed air strikes and launched cruise missiles against various sites across Iraq, including the Zaafaraniyah nuclear fabrication facility. Once *London* was back in the Gulf, Captain McClement had ensured she was fully integrated with the US Navy:

> If there was conflict the plan was for us to join up with the US Task Force and so we went and joined a US Navy Carrier Battle Group to perform a surveillance role.

While the Iraqi Navy remained out of the equation, Saddam had preserved his air force and also managed to field a number of Silkworm missile launchers on Iraq's coast. But they remained dormant and, by 20 January, the Iraqis were buckling under the American onslaught and asking for talks to discuss terms of access for the UN to weapons facilities.

Missile Guard

London returned to the UK in February 1993 and spent the next six months in home waters, undergoing maintenance, providing a platform for training warfare officers and visiting Falmouth and London.

During September she went through the usual pre-deployment operational sea training at Portland, which tested both ship and crew to the limit in all areas of warfare and damage control, arriving in the Adriatic the following month.

Savage civil wars raged across the Balkans, with Serbia sponsoring aggression in Croatia and Bosnia. British peace-keeping troops had been sent into Bosnia on behalf of the United Nations during the winter of 1992 and the Royal Navy had committed a land-based helicopter squadron to support them. In mid-January 1993 a British naval task group, led by the carrier HMS *Ark Royal*, sailed for the Adriatic to provide air support via her Sea Harrier strike-fighters. With her was the aviation support vessel RFA *Argus*, pressed into service as a makeshift assault ship carrying 105-mm guns and commando gunners on standby to be

A view of HMS *London*'s forecastle, taken in summer 1995 during work-up training at Portland just prior to her post-refit Adriatic deployment. She was the last warship to undergo intensive combat training at the Dorset naval base before its closure.

Nigel Andrews/active vision.

Training for the Adriatic included realistic mock fire-fighting exercises. *Nigel Andrews/active vision.*

inserted into the Bosnian battleground. The task group also provided an evacuation option if reinforcing the UK's peacekeepers failed. In addition, British warships had been helping to enforce a UN arms embargo against the warring factions since November 1992 and Serbia was also prohibited from importing oil.

The Serb fleet – three Koni Class frigates, ten Osa and Koncar class missile craft and four Sava and Heroj class submarines – was principally located in Kotor on the coast of Montenegro, and when *London* arrived in the Adriatic in late October her principal function was using powerful sensors to provide early warning of any Serb naval threat to the British carrier (by then HMS *Invincible*). Captain McClement placed his frigate just beyond Serbia's territorial waters and kept his crew at a high state of alert:

On picket duty off the coast everyone aboard carried their lifejackets at all times and always had their anti-flash clothing to hand. You also had to have gas masks available by your bunks or where you were working.

But not every navy was as vigilant about the potential threats, as Captain McClement saw, when *London* passed a warship from an allied nation:

I won't say what country the ship was from, but they were between the coast and us, and they were having a barbecue on the flight-deck and sunbathing. Seeing this, my people were asking me why on earth we were working in defence watches, six hours on and six hours off. I told them that the threat was very real... we could be targeted by a missile and must be ready at all times. I told one sailor: 'I am not prepared to write to your mum and tell her you are dead because I relaxed while my ship was within missile range.' On some patrols you were on the territorial limit and on others beyond it, but more often than not you were within range of shore-based weapons. A few days later we were locked up by a fire control system situated on the same stretch of coastline, possibly a missile system. They were probably carrying out a daily check on their fire control radar. After that incident I got no questions about why we were in defence watches and not sunbathing or having barbecues.

Winning the Sword of Peace

In November HMS *London* stood down to visit the Albanian port of Durres, the second British warship to call at the poverty-stricken former communist country since it had emerged from decades of self-imposed exile. The Foreign Office wanted *London*'s visit to reinforce the UK's efforts to build links with Albania. It was a task Captain McClement and the frigate's crew accepted willingly, despite Albania not being the top choice for their first run ashore:

> It was tough for my sailors because they had just completed three weeks on defence watches and straight afterwards they were going back to sea for two more weeks of it. This was their one run ashore in the Adriatic. On the advice of the local police and UK Consulate we had a 6pm curfew for the sailors if they went ashore.'

Petty Officer Martin Sykes was one of those who experienced the 'delights' of Durres:

> It was a Third World country and not a pleasant place at all, but you had to have some sympathy as they were just emerging from decades of dictatorial rule. As was the custom with Royal Navy port visits, we did what we could to help them out.

Captain McClement was extremely impressed with the manner in which his young sailors rose to the task:

> To their credit they didn't mind and did a lot of good work for the local community while we were in Durres. They scrubbed out a hospital's kitchen, mended catering equipment and made it usable again. They went to an orphanage that was in a dreadful state and helped clean it, decorate it, mended furniture and gave every chocolate bar we had onboard to the kids. In co-operation with charities and the UK government's Overseas Development Agency our helicopter flew blankets, clothes and other supplies up to villages in the mountains that had been snowed in. The helicopter flew off the back of the ship for eight hours a day over two days.

After his ship left Albania to return to picket ship duties, Captain McClement accepted an invitation to see what life was like on the ground with the British Army in Bosnia. In early December he travelled through a treacherous, snow-bound landscape to Gornji Vakuf, one of the more exposed British outposts. He later wrote of his experiences:

Clad in flash gear and overalls, *London*'s sailors man guns and signal lamps. *Iain Ballantyne.*

HMS *London*'s Ops Room sailors were trained to spot emerging threats before they could endanger the ship. *Nigel Andrews/active vision.*

> *Going through the town you have to wear your helmet, and if you are in an armoured vehicle you have to be inside it 'buttoned down'; you are not allowed to stop. If you have a problem with your vehicle you have to abandon it and everything inside, with the exception of your personal weapon, helmet and flak jacket, and get into the next vehicle in the convoy. While we were in the camp there was artillery fire going over our heads (not aimed at us), close small arms fire and one grenade![2]*

The *London*'s Commanding Officer returned from his trip up-country unharmed and the frigate was soon heading back to the UK where she was due to go into Devonport Dockyard for

During a mock combat exercise off the Dorset coast *London*'s bridge is at a high state of alert to spot in-coming 'air attacks'. As can be seen from the sign on the door, the tradition of naming parts of the ship after stops on the London underground, or boroughs of the British capital city, was carried on into the mid-1990s. *Nigel Andrews/active vision.*

her first refit. But before she paid off, Captain McClement held a dinner onboard for the ship's sponsor, Dame Mary Donaldson, who had launched her in October 1984, and other special guests. In a speech to the assembled dignitaries in *London*'s wardroom Captain McClement said:

> *Since* London *was commissioned in February 1987 she has steamed 266,000 miles, travelled as far north as Bear Island in the Arctic Circle, to Mombasa in the south, to Singapore in the East and to Florida in the west... During the 985 days spent at sea the Ship's Company consumed 50 miles of sausages, half a million tea bags, 10,000 gallons of custard and 330 tons of potatoes; I'm very glad I didn't have to foot that shopping bill!*

The Captain concluded:

> *Samuel Johnson's words sum up perfectly what I feel at the end of my time as* London*'s captain. He said: 'If a man is tired of London, he is tired of life, for in London there is all that life affords'.*[3]

The Final Commission

During the rededication service at Devonport Dockyard on 24 March 1995, Captain McClement returned to his old ship to see her new Commanding Officer, Commander Alan Brooks, receive the much-coveted Wilkinson Sword of Peace for the ship's humanitarian achievements during the visit to Albania.

That July *London* was the last warship to undergo operational sea training at Portland, before the Royal Navy's OST organization transferred to Devonport.

In September, before embarking on another Adriatic deployment, the *London* sailed up the Thames to visit the city after which she was named. There to meet her were Gordon Bruty, Graham Bramley and other veterans from cruiser *London*. Fifty years after the event, the veterans had managed to persuade the city of London to hold a special reception to honour their *London*'s part in the Second World War, and make up for the lack of a proper welcome home in 1945. Seventy-four veterans were present at a luncheon in Mansion House, which was presided over by the Duke of York, himself a serving naval officer.

Returning to Devonport, HMS *London* set sail for the Adriatic in October. Back in HMS *London* after eighteen months away learning to be a warfare officer, was Lieutenant Lee Abernethy. He found the British naval commitment off the Balkans remained at a high level:

> *The ship was in defence watches the whole time we were there, sitting on the 12 mile limit in a high state of alert. The Serbs' Galeb jet fighters were based somewhere north of Kotor and they used to regularly fly over the sea to come in to land at their airfield. We always watched them carefully to make sure they didn't come our way. They also flashed up their missile batteries on the coast every now and then.*

By November 1995, intervention by NATO air forces and a more forceful use of firepower by UN peacekeeping forces, together with the Croat faction gaining superiority over the Serbs on the battlefield, allied with the severe impact of sanctions on Serbia, persuaded the warring factions to sign the Dayton peace agreement. To enforce this agreement NATO deployed a large Implementation Force (IFOR) into Bosnia, which led to a change of plan for *London* and other British warships.

> *A spokesman at the Royal Navy's Fleet Headquarters said planned port visits to Barcelona, Malta and Majorca during the festive season have been abandoned. Now the aircraft carrier HMS* Illustrious, *her escort ship HMS* London *and support ships will steam just off the coast of the Balkans.*[4]

HMS *London* passes through the Corinth Canal in late 1998, on the way home from her final overseas deployment. *Guy Toremans.*

The Sea Harrier strike jets aboard *Illustrious* were needed as part of the air cover shield for NATO forces as they went into Bosnia, but once NATO troops were well established ashore, the British naval presence was wound down, with *London* ordered home in April 1996. Before leaving the Adriatic, the frigate paid another visit to Durres where her sailors carried out more good deeds in Europe's poorest country.

That summer HMS *London* was the last Royal Navy surface warship to do a surveillance patrol shadowing the Northern Fleet's summer manoeuvres in the Barents Sea. She encountered a brand new Oscar Class cruise missile submarine – the *Kursk*, which sank with all hands four years later when her own weapons exploded while she was submerged.

In May 1997, the *London* crossed the Atlantic for a seven month deployment that took her from Newfoundland to the Caribbean. In July she discovered an abandoned yacht drifting in the area of the so-called Bermuda Triangle. Sailors from *London* who went aboard the boat found 'clothes and personal belongings lying around, and even an open book on a bunk'.[5] Despite the area's dreaded reputation, *London* sailors, temporarily detached to crew the yacht, managed to escape the Bermuda Triangle and sail her to Puerto Rico where they rejoined their warship. Not long after this a disaster relief exercise in the West Indies became real when the warship ran into Hurricane Erica, but she made it back to Plymouth unharmed by December 1997.

Axed Before Her Time

If the seas could not claim her, the defence spending axe could. In the summer of 1998, the UK's newly elected Labour government unveiled its Strategic Defence Review. SDR laid out the proposed shape of the Royal Navy by the year 2015, which included two new 50,000 tons aircraft carriers, at least six 7,000 ton Type 45 air-defence destroyers and up to five 7,000 tons Astute Class hunter-killer submarines. Bigger, but fewer, than the ships they replaced, these vessels represented an ambitious leap in capability for the Royal Navy.

In the short term there were to be cutbacks that included retiring from service all the Type 22 (Batch 2) frigates (the Batch 1s had been sold to Brazil just over three years earlier).

With only eleven years on *London*'s clock it seemed a dreadful waste of a fine ship at a time when worldwide commitments from the Gulf to the Falklands were inflicting severe over-stretch on the British fleet. But, faced with a choice between running *London* and her sisters on, and investing scarce money in further refits, the Navy Board elected to pay them off. With no 4.5-inch gun to support ground troops in an era of littoral warfare and an obsolete anti-shipping missile system (Exocet) they were regarded as being of limited use without major modification. The Royal Navy was also facing a severe manning crisis. With recruitment having been wound down in the aftermath of the Cold War it was only slowly picking up, and in the meantime sailors from retired Type 22s could be used to fill manning gaps in other warships.

By the summer of 1999 HMS *London* would be paid-off and laid up at Fareham Creek in Hampshire, for sale to a foreign government or, if that failed, scrapping.

Her final overseas deployment was scheduled to begin in the autumn of 1998 and would take her to the Mediterranean and the Black Sea. She was ordered to leave a week early because of an escalating crisis in Albania where an economic disaster had sparked insurrection. *London* would stand by with other Royal Navy ships in the Adriatic to evacuate up to 200 British nationals. However, the evacuation plan was not implemented and so *London* resumed her programme, representing Britain in NATO's Exercise Dynamic Mix in the eastern Mediterranean in October before entering the Black Sea for a friendship visit to the Russian port of Novorossisk, twin city of the frigate's home port of Plymouth.

HMS *London* enters Plymouth Sound flying her paying-off pennant and greeted by two tugs firing water hoses. *M. Welsford.*

The final overseas deployment also included exercises with the navies of the Ukraine and Bulgaria and a visit to poverty-stricken Romania, where the ship's sailors again did their best to help out those less fortunate than themselves.

After returning to the UK in late January 1999, the *London*'s Commanding Officer, Commander Tom McBarnet, was asked to take his warship on a tour of six major British ports to raise the Royal Navy's profile at home as part of a recruitment drive. Arriving back from this last journey in late May, the *London*'s formal decommissioning took place on 11 June, with special guests including her Gulf War Commanding Officer, Iain Henderson, by then a Rear Admiral, and Dame Mary Donaldson.

De-stored by August, the ship made her way under tow to Fareham Creek, where she was put into mothballs.

Pinochet's Arrest Lets in Romania's Bid

A potential buyer for *London* and other ships in her class had been found in late 1998 – Chile. The South American country was looking to replace former Royal Navy County Class destroyers purchased in the 1980s.

> *The Chief of the Naval Staff was to fly to the UK in the last week of October that year to sign the contract.*[6]

The arrest for alleged war crimes of the ailing, elderly former Chilean dictator General Pinochet, while receiving treatment in a British hospital, scuppered the deal.

By the end of 2000, with Pinochet returned to his home country, the Chileans were showing interest in Type 22s again. But now they were requesting purchase of the Batch 3s, as they disliked the fact that the Batch 2s had no gun or effective anti-surface missile system. While talks dragged on without result – the British fleet would rather not sell its very capable Batch 3s earlier than it needed to – by October 2001 Romania had expressed an interest in buying *London* and *Coventry*, the latter due to pay off by the end of that year. By purchasing the pair, Romania would be achieving the quantum leap in naval capability it needed to qualify for membership of NATO. In April 2002, it was revealed that Romania was offering £106 million for the *London* and the *Coventry*.[7] The UK and Romanian governments were discussing the finer details of the deal, with the formal contract expected to be signed by the end of the year. Substantial modifications to the ships would probably include installation of a 76-mm gun and a new anti-shipping missile system.

The UK defence giant BAE SYSTEMS would work with the new owners to carry out the bulk of the work in Romania.

The proposed purchase of these two ships will provide the Romanian Navy with modern, powerful ships with an expected future life of 20-30 years.[8]

Provided the sale did not fall through, it was anticipated that by 2004 *London* would have completed her reactivation and modifications to join the Romanian fleet under a new name.

In early 2003 there was speculation that Romania might withdraw from the *London* and *Coventry* purchase, with Chile stepping in to buy one or both of the warships. Either way, the name *London* would now be free for use again in the British fleet.

Notes

1. Lieutenant Commander Lee Abernethy, in an interview with the author.
2. The personal papers of Rear Admiral Tim McClement.
3. ibid.
4. News report filed by the author, 18 December 1995, while working as Defence and Diplomatic Correspondent for the London Bureau of UK News.
5. *The Times*, 16 July 1997.
6. *WARSHIPS IFR* magazine, Oct/Nov 2000.
7. *WARSHIPS IFR* magazine, June/July 2002.
8. BAE *Systems' Response* news publication, April/May 2002.

Chapter Twenty-Two

21st CENTURY SUBMARINE HUNTER

In late 2018, it was revealed by the Royal Navy that there was to be another HMS *London*, a City Class (Type 26) frigate with her main function Anti-submarine Warfare (ASW). The wisdom of ordering a new class of ASW warships became ever more clear as the opening years of the 21st Century evolved into an era of great power rivalry, especially at sea.

For the next *London* is likely to find herself operating on the front line of a global East-West face off, expected to confront the undersea menace posed not only by a resurgent Russia but also counter Chinese maritime power projection around the world.

Confirmation of the new *London* came from Prime Minister Theresa May during the annual Lord Mayor's Banquet, which was in 2018 held on Monday, 12 November – the day after the UK commemorated the 100th anniversary of the end of the First World War.

How the next HMS *London* will look on deployment, carrying out a variety of frontline missions into the 21st Century. *Courtesy of BAE Systems.*

The P.M. said she was 'proud to be able to announce the naming of HMS *London* – one of our eight planned Type 26 frigates.' Mrs May added: 'As she upholds global stability, she will also bear the name of this great centre of trade and finance, reminding us all of the critical link between global stability and global prosperity.'[1]

* * *

It was in early 2010 that the then First Sea Lord, and former captain of the Type 22 frigate HMS *London*, Admiral Sir Mark Stanhope revealed in an interview with the author of this book, that a new kind of major warship would be known as the 'Type 26'.[2]

Admiral Stanhope's revelation related to the fruit of the UK's Future Surface Combatant (FSC) study, which, over several years, sought to devise a range of new vessels. In the interview the First Sea Lord said:

> It is no secret that the FSC programme is later than originally designed. I can tell you that the FSC C1 version is going to be called the Type 26 Combat Warship. One of the reasons we are calling the FSC a Combat Warship, is to broaden peoples' mindset. We want the FSC to deliver a range of effects and therefore we aren't necessarily starting off with the current mindset of a frigate combatant.

The design actually ended up being known as the Global Combat Ship (GCS), indicating high hopes of world-wide sales that were indeed fulfilled. It was chosen in June 2018 by the Australians (as the Hunter Class, with construction due to start in 2022) and the Canadians, in early 2019, as their Canadian Surface Combatant (CSC), with build starting in 2024.

The Royal Australian Navy (RAN) is to receive nine and the Royal Canadian Navy (RCN) fifteen of the vessels respectively, in bespoke variants, which, according to BAE Systems, the UK defence company spearheading the global project, supports 'greater operational, training and intelligence ties between the three nations.'[3]

* * *

With thirty-two Type 26 frigates in service with the RN, RAN and RCN, it will be ubiquitous on the high seas. It is possible the future *London* will operate alongside sister vessels from the RAN and RCN within joint British-Australian-Canadian task groups, likely centred on the new Queen Elizabeth Class carriers of the RN.

From the deck down the Type 26 variants are more or less the same, with BAE Systems Naval Ships in the UK, BAE Systems Maritime Australia and Irving Shipbuilding in Canada all working very closely. It is in the combat systems that they will differ markedly, with, for example, the Hunter Class and CSC expected to have an Aegis-based combat management system and Standard missiles for long-range air-defence tasks. The City Class of the Royal Navy will be fitted with the short-range Sea Ceptor Anti-air Missile (AAM), reflecting the fact that the Royal Navy already has six destroyers for long-range air defence while the Canadians have no destroyers and the Australians just three. *See below for more on City Class weapons and sensors.*

The exact fit for the different Type 26s is still being settled, but all possess vertical-launch systems that enable them to embark a variety of weapons for air-defence, ASW and land-attack roles. Canada's first Type 26 will be delivered in the early 2030s, while the RAN will commission its Type 26s from 2031 onwards.

A CGI artist's depiction of the three variants of the Global Combat Ship (Type 26) warship at sea together, with the City Class (UK) nearest, Hunter Class (Australia) middle and Canadian Surface Combatant furthest. *Stephen Mchugh/BAE Systems.*

A Bid to Unite the UK in Troubled Times

The Type 26 project will not just unite the navies of nations that were part of the British Empire when the cruiser *London* sailed the seas. It will also aim to knit the four home nations of the United Kingdom together in a time when the bonds that bind them are being strained.

According to the nationalist government in Edinburgh, Brexit has unfairly ripped the Scots out of the European Union (EU) against their will. The ruling Scottish National Party (SNP) is at the time of writing expressing its fury with the current settlement by pushing relentlessly to break Scotland away from the British political and economic union. Meanwhile, there is turmoil over a so-called border in the Irish Sea – imposed by new trade regulations after the UK left the EU. It is seen by Ulster Unionists as a dirty trick to try and remove Northern Ireland from the United Kingdom.

As if to emphasise the unifying power of the Royal Navy – always a proudly British organisation – the Type 26 frigates are named after cities in each of the UK's constituent nations, not least the capitals of Scotland, Wales, Northern Ireland and England. The first Type 26 built for the Royal Navy is the future HMS *Glasgow*, followed by *Cardiff*, *Belfast*, *Birmingham*, *Sheffield*, *Newcastle*, *London* and *Edinburgh*.

Construction began in the summer of 2018 and by July 2020 the last of 57 'segments', as the Royal Navy described it, was being created to complete 'the 8,000-tonne jigsaw which is HMS *Glasgow*.'[4]

In fact, the first three ships had all been ordered by then – with *Cardiff* and *Belfast* laid down – and with the overall price tag for the initial trio £3.7 billion.[5]

When the UK built sixteen Type 23 frigates between 1985 and 1999 the cost of the first (*Norfolk*) was £135.449 million, with the last (*St Albans*) £96 million.[6] Naturally costs rise with time due to inflation and the Type 26 is a larger, more complex and powerful vessel than the Type 23. It was confirmed by the Integrated Review of defence and foreign policy (IR2021) published in March 2021 that the remaining five Type 26s would be ordered.

In the IR2021 Command Paper, the MoD outlined investment in not only the eight Type 26 frigates, but also five Type 31s and five future Type 32 frigates, the latter still in the conceptualisation phase.

The Type 31 is to be a less complex general-purpose frigate, with construction beginning at Rosyth Dockyard in September 2021 and estimated to cost the UK taxpayer £250 million per ship. There are likely to be additional costs due to new weapons and sensors on top of those in the basic specification. The Inspiration Class (as the Type 31s are also known) are being built by Babcock, breaking the BAE Systems monopoly on UK frigate construction, though keeping such work the preserve of Scottish yards.

The IR2021 Command Paper outlined a role for the Type 26 and Type 31, much like that performed by the cruiser *London* between the two world wars of the 20th Century. According to the Command Paper investment in the Type 26s and Type 31s 'will provide the modern vessels required by the Royal Navy to protect our territorial waters and the global shipping routes on which our economy relies.'[7] Mrs May's successor as P.M., Boris Johnson, explained in Parliament, during a debate following his outlining of the broad scope of the 2021 review:

By the end of this decade, we will have 24 frigates, as opposed to the 15 today.[8]

But he misspoke, for the RN in 2021 had 13 frigates and it was later clarified by a Ministry of Defence (MoD) spokesperson that what the P.M. actually meant was the UK would have 24 destroyers and frigates.

Were the P.M.'s timetable adhered to it would entail a radical acceleration and expansion of the build programmes, but, realistically, the new HMS *London* cannot be expected to start construction until the end of the 2020s.

She probably won't be commissioned until the mid-to-late 2030s and so there may be young sailors in her crew who were not even born when Mrs May made her Lord Mayor's Banquet announcement.

Naming a ship so far in advance of any steel even being cut is a tactic the Royal Navy has used before in recent times, including by then First Sea Lord Admiral Alan West in the early 2000s. He ensured the two future aircraft carriers were named *Queen Elizabeth* and *Prince of Wales*.

In that way it became hard for politicians to cancel them without suffering immense political damage. While people may not have cared when the Type 22 frigate's name was changed from *Bloodhound* to *London* (*see p.169*), cancelling warships carrying the name of the sovereign and the future king, or even that of a city, would be a different matter.

It was interesting to note that Lord West of Spithead was two decades later not averse to changing the names of warships.

In April 2021 he wrote to '*The Daily Telegraph*' newspaper suggesting that it would be a fitting tribute to the recently deceased HRH Prince Philip, the Duke of Edinburgh – a career naval officer before giving it up to serve at Her Majesty The Queen's side – to rename one of the Type 26s HMS *Duke of Edinburgh*, though which one was not revealed. Lord West suggested a future *Duke of Edinburgh* 'would be able to escort HMS *Queen Elizabeth*'.[9]

Should such a thing come to pass, and *London* was the one to have her named changed, there is no doubt – despite the cause being worthy – there would be dismay in some circles within England's capital.

Changing the names of any other ships in the class would be seen as terrible insult to the home nations, achieving the exact opposite of naming them after cities, unless the future *Edinburgh* became *Duke Of Edinburgh*. Quite what the notably attention-averse Duke of Edinburgh would have thought to the ship re-naming notion we can only guess, but he did know a previous a *London*.

In 1964, at the controls of Whirlwind helicopter, he flew himself aboard the County Class *London*, which was sailing off the UK with a task group of anti-submarine frigates. The Duke was given a tour of the destroyer, including visiting the bridge and inspecting the Seaslug missile launcher, also chatting with her sailors who later gave him three cheers before he departed.

Returning to HMS *London* in 1965, the Duke displayed his talent for humorous observation when complimenting the accuracy of the Seacat missile aimer during a test firing. *As related on p.156-7.*

In May 2021, it was pointed out by a British naval historian writing to 'The Sunday Telegraph', reference the possibility of constructing a new Royal Yacht or warship named *Duke of Edinburgh*, that it would be far from straight forward.

Richard Johnstone-Bryden pointed out one of the new Type 26s is already to be named *Edinburgh*, and this could present a 'potential complication'.[10] Lord West in a letter published in the same edition of 'The Sunday Telegraph' wrote that he would be 'delighted to see a new ship called HMS *Prince Philip*' which could offer a number of benefits to the UK 'as well as righting a historic wrong: the loss of a royal yacht for the Queen's use.'[11]

Not long after the forward section of the first City Class frigate is rolled out on the Clyde the aft section is very carefully joined with it, to make a whole vessel at the Govan shipyard of BAE Systems. *Courtesy of BAE Systems Maritime – Naval Ships.*

Apparently, according to an earlier reports in *The Daily Telegraph*, based on UK Government sources, there were plans to set aside £200 million for a new Royal Yacht named in honour of the Duke of Edinburgh and operated by the Royal Navy. This would allay fears of a City Class frigate being renamed.

Picking up the Drumbeat

The last frigate actually completed in a UK shipyard was the Duke Class (Type 23) frigate HMS *St Albans*, in 2001. With first steel cutting for the *Glasgow* not taking place until the summer of 2017, it was a gap unprecedented in British naval history, silencing a drumbeat of frigate construction that had been steady for centuries.

Between 1980 and 2001, the UK actually completed 35 frigates and destroyers. After delivering six Daring Class destroyers (2008 – 2013) the UK's warship construction industry, such as it was, focussed on constructing new aircraft carriers, five patrol vessels and nuclear-powered attack submarines.

Once the drumbeat was picked up again, the job of building the Type 26 began in earnest at BAE Systems' Govan and Scotstoun yards on the River Clyde. With *Glasgow*'s build underway, first steel was cut for the future *Cardiff* in August 2019. Amid criticism in some quarters that the pace of construction was somewhat slow, in late February 2021 Lord West tabled written questions in the House of Lords that included asking about the 'planned launch date of that frigate [*Glasgow*]; and when it is expected to be accepted into service by the Royal Navy.'

Defence minister Baroness Goldie of Bishopton replied on 9 March by explaining: 'The £3.7 billion contract to manufacture the first batch of three Type 26 frigates was awarded in June 2017 and steel cut on the first ship, HMS *Glasgow*, in July 2017.'[12] She carried on: 'I have interpreted the Noble Lord's use of the term "launch date" to mean when HMS Glasgow will be in the water. On current plans, HMS *Glasgow* will be floated-up in BAE System's Govan shipyard in the second half of 2022 and then transferred to its Scotstoun shipyard. There, she will be fitted-out with complex weapons and other systems. After completion of sea trials and Royal Navy training and preparations, she is currently expected to enter service in 2027.'

Lord West did manage to tease from Baroness Goldie that the next two Type 26s – *Cardiff* and *Belfast* – will not enter service until 'the late 2020s.' When he pressed her to confirm the situation on ordering more Type 26s she stonewalled the former First Sea Lord by explaining: 'As the contract award for the Batch 2 Type 26 frigates is subject to commercial negotiation and Her Majesty's Treasury approval, I am unable at this stage to provide a more precise timescale.'[13] Baroness Goldie pointed out that, with the Type 26 construction already underway and the Type 31s starting in 2021 there would 'for the first time in 30 years two classes of frigates simultaneously under construction in UK shipyards.'[14]

In the 'Defence and Security Industrial Strategy' paper of IR2021, it was mentioned that there were to be eight Type 26s built, but there was no confirmation of the second batch being ordered.

The IR2021 also did not dispel fears over the Royal Navy's frigate force lacking the kind of critical mass needed to handle combat with the Russians or the Chinese whom the new HMS *London* may face on escort missions.

The shape of things to come: The Royal Navy strike carrier HMS *Queen Elizabeth* **sails out of Portsmouth in May 2021, to begin her first global deployment. Future UK Carrier Strike Groups (CSG) led by** *Queen Elizabeth* **will likely include the future HMS** *London.*
Gary Davies/Maritime Photographic.

And one day a lack of warships could have serious consequences. While it is likely the new *London* will be operating in a task force alongside warships from European navies and the American, Japanese Australian, Canadian and Indian fleets, the old maxim that the best protection is from your own Navy holds true.

* * *

Whether in conjunction with allies, or on her own, what can the future HMS *London* offer within a Carrier Strike Group (CSG) or while acting in some other role?

At 151 metres long, 7,000 tons displacement and with a crew of 157, plus the ability to accommodate more than 50 additional personnel (such as Royal Marines), the new frigates will have a range in excess of 7,000 miles, with a top speed of more than 26 knots. Similar to the *London* of the Cold War, the City Class warships are destined to be based at Devonport Naval Base in Plymouth, where they are replacing the eight Anti-submarine Warfare (ASW) configured Type 23s.

Like those warships the future *London* and her sisters will have towed array sonar to enhance sub-hunting capabilities, along with an embarked helicopter fitted with sonar, plus armed with missiles and torpedoes.

What is described as a secondary hangar will enable the Type 26 to host Unmanned Aerial Vehicles (UAVs) to put eyes-in-the-sky and spot potential threats from further away. A flexible mission bay will be able to accommodate and launch drones in addition to small boats.

Armed with the Sea Ceptor AAM and a Vertical-launch System (VLS) capable of launching both land attack and anti-submarine weapons, the City Class will have a 5-inch medium calibre gun.

The new *London* will be able to: launch raiding groups of Royal Marines, conduct surveillance using drones or her sensors; provide Naval Gunfire Support (NGS) for troops fighting ashore; protect an aircraft carrier from air attack; evacuate British nationals from war zones and deliver disaster relief and humanitarian aid.

The aforementioned tasks have been the stock-in-trade of the Royal Navy's cruisers and frigates for decades but it is the anti-submarine role the future *London* will likely be most heavily engaged in as the world rewinds to a familiar kind of confrontation.

Post-Cold War Amity put to the Sword

Amid all the joy of the Cold War ending in the Barents Sea during the summer of 1991, the frosty enmities of more than 40 years were thawed by plenty of vodka and toasts to peace and new friendships.

As we have seen earlier in this book the Type 22 frigate HMS *London*'s visits to Murmansk and Archangel were a triumph. The sailors of both the Russian and the British navies endured some monster hangovers to draw a line under a dark and dangerous post-war era.

The end of the Cold War between the Royal Navy and Soviet Navy in the Barents Sea, August 1991. HMS *London* is greeted in friendship by the Soviet Navy warship *Gromky* and saluted by two Sukhoi fighter jets. *London*'s Lynx flies shotgun astern, with the tanker RFA *Tidespring* in the background. The author of this book is among the figures waving to the *Gromky* from the bridge roof of HMS *London* and helped the artist with research for composing this work.

Painting by Ross Watton © 2013. For more on the work of Ross Watton visit https://www.rosswatton.com

Who, back in the summer of 1991, would have thought that one day the frostiness between Russia and Britain would return with a vengeance and there would be a new face off?

It was assumed that Russia would follow the path to a Western-style democracy. Just as the Chinese regime has adopted a kind of totalitarianism that mixes naked capitalism with old school communist-style repression of political and civil freedoms – while competing militarily with the West – so also has the former Soviet Union. Under the leadership of former KGB spy President Vladimir Putin, it has not only pursued so-called hybrid warfare – embracing assassinations on foreign soil, cyber-attacks and use of mercenaries in various conflicts – but also beefed up its conventional and nuclear forces.

In April 2021, during his Presidential Address to the Federal Assembly in Moscow, Putin boasted about new naval firepower, including long-range hypersonic cruise missiles and a massive nuclear-tipped drone called the Poseidon. In essence a huge high-speed torpedo, the latter can be launched by large submarines to devastate an enemy's ports.

Putin warned the West that Russia will not tolerate 'provocations that threaten the core interests of our security' and that opponents 'will regret what they have done in a way they have not regretted anything for a long time.' Putin added: 'But I hope that no one will think about crossing the "red line" with regard to Russia. We ourselves will determine in each specific case where it will be drawn.'[15]

Yet, even in 1991 there were signs of mistrust lingering under the surface when the British frigate's senior ratings, including CPO Yorkie Cunningham, went aboard the Russian frigate *Gromky* at Murmansk. The *London* group also included the author of this book, in a journalistic capacity, who similarly found the Russians touchy about certain topics despite the general bonhomie. 'We realised what a façade it all was,' CPO Cunningham would later reflect.

We were there to be sociable, but also to take notes on what their ships were about, in case there was ever another [hardliners'] coup that made us into foes again. Their house-keeping aboard Gromky *was far from impressive. I remember they opened a locker full of fire-fighting gear which was all green and grimy, not well looked after at all. You'd get into big trouble on the* London *if you did that. In the Ops Room I asked if I could see how the sonar worked and our guide took a look at my sonar badges and gave me a polite 'no'. On the upper deck it was quite rough underfoot and I realised that they didn't rub the old paint down – they just painted over the top.*

That less than professional approach reaped terrible consequences for other Russian sailors in August 2000 when the Oscar II Class nuclear-powered submarine *Kursk* sank in the Barents Sea. The *Kursk* met her end thanks to a massive weapons compartment explosion caused by poor maintenance and a lack of training in how to handle torpedoes safely.

In the aftermath of that incident, President Putin vowed that his nation's navy would clean up its act and be modernised to again become the global standard bearer of Russian power and influence.

Cold War warriors like surviving Oscar Class guided-missile submarines (SSGNs) were taken in hand for regeneration. Russia also built new nuclear-powered attack boats and ballistic missile submarines (SSBNs), along with special purpose craft capable of carrying the new nuclear-tipped drones and/or sending down mini submarines to cut the West's internet cables. Moscow commenced a programme to construct up to 18 new Improved Kilo Class diesel-electric attack submarines (SSKs) along with cruise-missile armed corvettes and frigates.

Today's Russian Navy has shown itself to still be a force to be reckoned with, via cruise missile bombardments of targets in Syria by those SSKs and also surface warships – propping up the regime of the dictator Bashar al-Assad by pummelling rebel forces. A 2019 global circumnavigation by the new frigate RFS *Admiral Gorshkov* – named after the legendary leader of the Soviet Navy during the Cold War – was another demonstration of reach and geo-political influence.

Such a turn of events would have seemed incredible in 1991 as the Type 22 frigate *London*'s sailors observed the wreckage of the Soviet Navy strewn along the shores of the Murmansk Inlet. For years afterwards the Russian Navy all but disappeared from the world's oceans.

But three decades on from *London*'s visit to Murmansk and Archangel the Russians are displaying a gritty determination to revive their naval power, along with its reach and presence on the oceans. A repeat invite for a Royal Navy warship to process down the river from the Kola Bay to Murmansk is unlikely any time soon, for Moscow obviously again views the UK as a foe. For its part the UK Govt stated in its IR2021 'Global Britain in a competitive age' paper that 'Russia remains the most acute threat to our [the UK's] security.'[16]

Following two joint patrols by the US Navy's warships and British frigates in the Barents Sea in 2020 – the first such deployments since the end of the Cold War – the Royal Navy revealed in March 2021 that it was preparing to regularly commit a frigate to patrols in Arctic waters. The aim is to challenge Moscow's ambition to claim the High North as *mare nostrum*. It was likely to be a joint force of UK, Norwegian and American vessels, with possible participation of other NATO navies.

Depending on how long relations between the West and Russia remain in deep freeze this time, one day the new HMS *London* could well be sailing in a hostile Barents. She and her crew will likely be locked into what was described in the Hollywood movie version of Tom Clancy's 'Hunt for Red October' as 'a war with no battles, no monuments...' We will pray that it remains so and never turns hot, though events during the Ukraine War do not augur well.

For with Russia aspiring to rival and even exceed Royal Navy – whose new aircraft carrier it describes as 'just a convenient, large maritime target'[17] – encounters between a future *London* and Russian vessels will not produce an exchange of cheers like they did in August 1991.

And, while making life uncomfortable for NATO and giving the American navy pause for thought, Russia has also been joining forces with the Chinese.

The UK and the West are simultaneously being confronted with the rise of Chinese maritime power in Indo-Asia-Pacific.

Both Russia and China are also keen to exploit the possibilities of new trade routes across the roof of the world made possible by global warming.

When the last HMS *London* was on patrol in the 1990s, the People's Liberation Army Navy (PLAN) of China was large but operated obsolete vessels and was not much more than a coastal defence force. Today it is on the cusp of true global power projection. To achieve maritime pre-eminence, China was by 2021 building 20 destroyers (with 26 already in service) and already had close to 100 modern corvettes and frigates in commission with more to come. The People's Liberation Army Navy (PLAN) had become the world's largest navy if not the most powerful or capable (that honour still went to the US Navy). China's

In this CGI artist's impression of a City Class (Type 26) frigate on patrol, a Merlin anti-submarine helicopter is landing on the warship's flight-deck. *Courtesy of BAE Systems.*

submarine force is composed of 66 vessels, including nuclear-powered and conventional attack boats and those will be exerting a presence in the most tense area of confrontation – the South China Sea – where the future *London* may also soon sail.

Jack of All Trades

The opponents Type 26 frigate *London* will face are likely to be formidably armed and numerous, contrasting with the small numbers of British warships that are, in general, comparatively lightly armed.

Thanks to a succession of defence cuts since 1990, today's Royal Navy is half the size it was in 1991. When Type 22 *London* sailed to Russia the RN could boast 13 destroyers and 37 frigates, 39 mine-hunters and mine-sweepers, 13 Offshore Patrol Vessels (OPVs), along with 12 diesel-electric patrol submarines, 17 nuclear-powered attack submarines and four SSBNs.

Today the RN has half a dozen destroyers, will soon have only 11 frigates, while also operating just eight OPVs, six SSNs, three active SSBNs plus will, within a few years, have no mine warfare vessels.

Yet it boasts two massive aircraft carriers and is setting up forward operating bases around the world, investing in three classes of frigates, new mine warfare capabilities focussed on deploying drones and proposing a new (Type 83) destroyer.

In the meantime, despite being so small, the Royal Navy's portfolio of missions is as broad as ever, and as a force multiplier against the Russians and Chinese, a lot of faith is being put in the aforementioned drone technology arming vessels like the future *London*. But it is immature as an effective weapon of war.

It also offers the risk of an enemy being able to detect remote-control signals and other emissions from mother ships and use them to destroy those vessels and neutralise the drones. Warships operating as stealthily as possible while being able to hit hard and at long range with missiles will count too in any future war involving *London*.

A navy as lean as today's RN arguably has too few people and ships to fall back on if it suffers battle damage, vessels needing major repairs to defects or in extended refit, or is suddenly faced with multiple crises. Therefore a major aim for the Type 26 will be maximising days per year on front line operations. This is likely to rely on forward-basing in trouble zones such as the Arabian Gulf and Asia-Pacific with the future *London*, for example, deployed for years to some base east of Suez. Fresh crews will be flown out from the UK and essential ship maintenance carried out at a local dockyard.

It has been pioneered in recent years by the Type 23 frigate *Montrose*, which has operated from Bahrain on maritime security duties to deter the Iranians from interfering with UK-flagged tankers in the Strait of Hormuz, a duty familiar to the Cold War era *London*. Forward basing is also a throwback to the days of the cruiser *London* in the late 1940s when she was sailing out of Singapore or Shanghai.

*   *   *

Warships and their crews must always be at the core of what a Navy does – and no drone can offer the same spectrum of benefits to the UK. For warships, along with their sailors and embarked Royal Marines, provide moral and legal authority in lawless seas. They also act as powerfully persuasive diplomats during foreign port visits, making friends and influencing people in time of peace.

While on deployment they can offer a means to deter an enemy from an act of war, preventing a situation from escalating into something uncontrollable via their mere presence on the high seas. They can also, of course, rescue UK nationals from war zones if need be, plus deliver humanitarian aid and disaster relief ashore.

Past HMS *Londons* have performed all those missions, as the preceding pages amply illustrate. The future *London* will be able to do the same, even if she is a hard-pushed part of the smallest British navy for centuries.

New Navy, Old Foe for the Type 22

Meanwhile, more than thirty years since the Type 22 frigate HMS *London* was first commissioned into service the same warship is back in the game of deterring the Russians. As *ROS Regina Maria*[18] she has patrolled the Black Sea and eastern Mediterranean in NATO task groups.

In July 2020 ROS *Regina Maria*, was part of Standing NATO Maritime Group 2 (SNMG2), alongside the Spanish frigate ESPS *Alvaro de Bazan* and Turkish frigate TCG *Yildirim*, participating in Exercise Sea Breeze, which was jointly organised by the Ukraine and USA.

This tested participants in a range of warfare skills, including air-defence and Anti-submarine Warfare (ASW). ROS *Regina Maria* could play only a limited part, as despite extensive work to prepare her for Romanian service she has no missile systems. Her current

The Romanian frigate ROS *Regina Maria* **(ex-HMS** *London***) in the Bosphorus, while on a NATO deployment.** *Cem Devrim Yaylali.*

armament consists of a 76mm gun, a pair of 30mm cannons and two triple-tube torpedo launchers, while she can also embark and operate a Puma helicopter.

The £116 million paid for the ships, ex-*London* along with the ex-*Coventry* (renamed ROS *Regele Ferdinand*), aroused great controversy in Romania in light of them being transferred without anti-air or anti-surface missile systems, or certain key sensors and Electronic Warfare (EW) capabilities. All of the latter had been removed when the ships were mothballed at Fareham Creek.

It has been suggested that the Royal Netherlands Navy (RNLN) could have sold the Romanians ships with missile systems for a lot less, due to the Dutch vessels not being put into mothballs first, but rather turned over 'hot'.

In 2006, a year after the reactivated former Type 22 frigates were transferred to Romania, it was proposed that providing them with replacement missile systems and sensors could be done for a further £250 million.[19] It was clearly a price Romania was not prepared to pay.

Despite her modest capabilities, *Regina Maria* is today flagship of Romania's presence in NATO, in a part of the world where the most powerful regional naval forces are those of Turkey and the Black Sea Fleet (BSF) of Russia.

Bulgaria and Romania, while NATO members, maintain only modest fighting fleets for they may be shy of arousing too much ire from Russia, bearing in mind their many years under the Soviet yoke during the Cold war.

The regular entry of powerful warships from NATO extra-regional navies, such as the those of the UK, USA, Canada and Spain or France, makes up for what Romania and Bulgaria lack.

Just contributing *Regina Maria* to SNMG2, along with her sister ship or a Bulgarian frigate, shows that the new Black Sea NATO nations are willing to do their bit and at least participate in warfare exercises.

Ukraine, which likewise had weak naval forces, hankered after joining the West's community of nations and wanted to become a member of NATO. That was a move that could provoke further military action by Russia or its proxies in eastern Ukraine. Russia had made it clear any move by NATO to move further east towards its borders would prompt a stern response. Despite that, Ukraine looked to the Alliance for a measure of security, which it provided via NATO warships patrolling international waters of the Black Sea.

But, never mind the Russian threat, in late 2020 it was revealed that *Regina Maria* faced an formidable internal enemy that caused her to be withdrawn from a deployment with SNMG2. It was while she was participating in the Turkish-led Exercise Mavi Balina in the eastern Mediterranean that the frigate suffered a COVID-19 outbreak. On returning home to Constanta, the frigate and her crew were immediately put into quarantine, with any sailors requiring medical attention taken to a hospital ashore. The frigate was still alongside in Constanta in February 2021.

By mid-March 2021, with her anti-COVID measures well in place, ROS *Regina Maria* had rejoined SNMG2 to participate in NATO maritime security patrols and also lead the way in an exercise organised by Romania called Sea Shield. Bearing in mind the Russian habit of harassing NATO warships with jet fighters, Sea Shield schooled participating vessels in air defence procedures.

On the agenda for *Regina Maria* and other warships of SNMG2 was a visit to Poti in Georgia, a small country which, like Ukraine, was at one time part of the Soviet Union. In 2008 defiantly independent Georgia had suffered a Russian invasion, with Moscow permanently occupying 20 per cent of its territory. It was a foretaste of the Crimean annexation of 2014, which saw Ukraine losing the naval base city of Sevastopol and its host peninsula to Russia. Like the Ukrainians, the Georgians wanted to NATO naval units like SNMG2 to stage defence diplomacy visits, in lieu of Alliance membership.

NATO warships on patrol on the Black Sea, early April 2021 with the Turkish frigate TCG *Kemalreis* (leading) and ROS *Regina Maria* (ex-HMS *London*) turning to starboard (middle) and the Turkish frigate TCG *Yavuz* (middle distance). *NATO SPAN L. Pons Miles.*

In the spring of 2021 war between Kiev and Moscow seemed entirely possible. Russian tanks, infantry, armour and artillery massed on the border to possibly invade eastern Ukraine, which, since 2014, had seen its Donetsk and Luhansk regions plunged into conflict by separatists Russia backed with its military muscle. As the dark clouds of new conflict gathered in mid-April 2021 the Russian Navy warned American warships to stay out of the Black Sea and staged missile firing exercises in waters off the Crimea. ROS *Regina Maria* had meanwhile stood down from duties with SNMG2 and was in port at Constanta, with her crew wondering if they would be called back from shore leave at short notice.

Yet, even as Putin was issuing his Moscow warning about a 'red line' to the West and NATO, the Russian troops were being pulled back from the border with Ukraine while the Black Sea Fleet's warships returned to port following their exercises. Had things turned ugly, and *Regina Maria* dared to venture out without protection from missile-armed ships of other NATO nations, she would likely only have showed the world how to die valiantly.

As that Ukraine and Black Sea crisis abated, at Govan on the Clyde a major milestone on the road to a new *London* was passed, with the forward section of the first City Class ship rolled out of the build hall of the BAE Systems shipyard.

It took 90 minutes for it be manoeuvred into position on the hardstand and was the first time the sheer scale of the cruiser-sized Type 26 had been revealed to the public. It was an impressive foretaste of the future HMS *London*. In the autumn of 2021, half a dozen sailors were drafted to the future HMS *Glasgow*, to start the process of breathing life into the lead vessel of the class.

The shape and size of the future *London* revealed: In April 2021 the forward section of the first City Class (Type 26) frigate, *Glasgow*, is rolled out at the Govan shipyard of BAE Systems on the Clyde. The next *London* will be the seventh Type 26. *John Linton/courtesy of BAES Maritime – Naval Ships.*

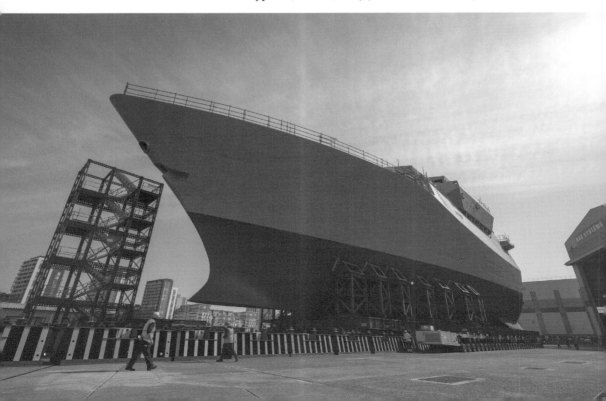

Notes

1. P.M. speech to the Lord Mayor's Banquet: 12 November 2018 – GOV.UK https://www.gov.uk/government/speeches/pm-speech-to-the-lord-mayors-banquet-12-november-2018
2. Admiral Sir Mark Stanhope, interview with Iain Ballantyne, published *WARSHIPS International Fleet Review*, March 2010 edition.
3. 'HMS Glasgow Takes a bow – Forward Section of First Type 26 is Rolled out', press release from BAE Systems, 19 April 2021.
4. 'Work Starts on Final Piece of HMS Glasgow's Gigantic "Jigsaw",' press release from Royal Navy, 23 July 2020.
5. BAE Systems.
6. Cost of Type 23s, House of Commons Hansard Written Answers 5 July 20001
   https://publications.parliament.uk/pa/cm200102/cmhansrd/vo010705/text/10705w05.htm
7. Command Paper, 'Defence in a competitive age', published March 2021.
8. Hansard, Boris Johnson during Integrated review debate, 16 March 2021.
9. Admiral Lord West of Spithead, in a letter to *The Daily Telegraph*, 16 April 2021.
10. Richard Johnstone-Bryden, in a letter to *The Sunday Telegraph*, 9 May 2021.
11. Lord West in a letter to *The Sunday Telegraph*, 9 May 2021.
12. Written questions, answers and statements – UK Parliament, 9 March 2021.
    https://questions-statements.parliament.uk/written-questions/detail/2021-02-23/HL13634
13. Baroness Goldie, answer to Lord West, 17 March 2021
    https://members.parliament.uk/member/3834/writtenquestions
14. Ibid.
15. President Putin during his state of the nation address in Moscow, 21 April.
16. 'Global Britain in a competitive age', IR2021 global vision paper, published by the UK Govt March 2021.
17. BBC News online report, 'Russia says news UK aircraft carrier "a convenient target", 29 June 2017.
18. The new name *ROS Regina Maria* honours the last Queen of Romania, a grand-daughter of Queen Victoria, who in the 1890s married the future King Ferdinand of Romania, himself honoured in the fleet by the ex-HMS *Coventry* renamed ROS *Regele Ferdinand*.
19. 'We paid three times too much for UK frigates, Romania says,' report in *The Guardian*, 13 June 2006 https://www.theguardian.com/uk/2006/jun/13/armstrade.bae

EPILOGUE
HOT WAR IN EUROPE

In February 2022 Vladimir Putin ordered a blitzkrieg of the Ukraine by 190,000 troops, using thousands of tanks and armoured vehicles supported by hundreds of strike jets and numerous missile-firing naval vessels.

If that was not bad enough, the Russian leader also threatened deployment of nuclear weapons against NATO nations to deter them from making any direct military intervention. The day of the invasion saw Putin issuing a threat to the West via a TV address broadcast to the world. He warned that anyone 'who would consider interfering from the outside…will face consequences greater than any you have faced in history.'[1] Putin was practicing what he had previously referred to as 'nuclear restraint.'[2]

The Ukrainians fought back with immense courage against the Russian invaders trying to seize key cities and capture major Black Sea ports.

Dismayed by his 'special military operation'[3] faltering within days, due to stubborn resistance by people he claimed needed liberating from a government of 'neo-Nazis',[4] the Russian leader arranged a televised conference with his defence minister and head of the armed forces.

Also stung by almost universal condemnation, and infuriated at stiff economic sanctions imposed on his regime by Western nations, Putin announced he was placing Russia's nuclear forces into what he called a 'special regime of combat duty.'[5] Having brought Russia's deterrent to a higher level of readiness, a few days later nuclear-armed submarines and surface warships were deployed into the Barents Sea to test their abilities to strike NATO targets. These were the same waters where the *London* and *Gromky* made their rendezvous in the cause of peace all those years ago. It was further conclusive proof that the bad old days had returned.

Putin warned NATO that even an attempt to establish a no-fly zone over Ukraine – which leading figures in the embattled country demanded should be imposed to stop the bombardment of its cities and killing of civilians – would be considered an act of war.

That too was a case of nuclear blackmail. For Russia's 'State Policy on Nuclear Deterrence' (published in June 2020) indicates that even conventional attacks by a foe against Russia or its strategic interests 'when the very existence of the state is in jeopardy' could trigger the use of atomic weapons. The same line of thought ran through Putin's invasion day justification for the assault on Ukraine. He suggested that 'in territories adjacent to Russia, which I have to note is our historical land, a hostile "anti-Russia" is taking shape.' He added that 'it [Ukraine] is doing everything to attract NATO armed forces and obtain cutting-edge weapons.

'For the United States and its allies, it is a policy of containing Russia, with obvious geopolitical dividends. For our country, it is a matter of life and death, a matter of our historical future as a nation. This is not an exaggeration; this is a fact. It is not only a very real threat to our interests but to the very existence of our state and to its sovereignty. It is the red line which we have spoken about on numerous occasions. They [the Ukrainians] have crossed it.'[6]

In face of Russia's relentless air and artillery assault on his nation, it was understandable that Ukraine's President Volodymyr Zelenskyy and others would demand a no-fly zone

ROS *Regina Maria* (ex-HMS *London*) seen from the forecastle of the Spanish frigate ESPS *Mendez Nunez*, during her time with Standing NATO Maritime Group 2 (SNMG) on patrol in the Black Sea, March 2021. A year later the Black Sea had become a war zone. *NATO SPAN L. Pons Miles.*

from NATO. But the Alliance's political and military leaders were clear that such an action would inevitably lead to a clash with Russia that could soon see events escalate into a broader war and the risk of tactical, or even strategic, nuclear weapons being used.

Amid all the razor's edge brinkmanship, there was no sign of ROS *Regina Maria* at sea to counter Russian aggression.

Following her brush with COVID-19, the frigate had deployed during the summer and autumn of 2021, taking part in NATO exercises in both the Black Sea and the Mediterranean, including with the French frigate FS *Auvergne* at year's end. But, due to the lack of suitable armament to defend herself properly – and without better-armed NATO warships as back-up – the ex-*London* declined to venture into the hot zone of the north-western Black Sea where the thunder of war resounded.

The Russian cruiser RFS *Moskva*, supported by a corvette and at least one Su-24 strike jet, attacked Snake Island, which is located close to Romanian territorial waters off the Danube delta. The *Moskva* radio'd the island's garrison of 13 Ukrainian border guards, suggesting they should surrender. This prompted the Ukrainians to declare: 'Russian Warship, go f**k yourself!' After the bombardment they did, however, capitulate.

Meanwhile, anti-shipping missiles and shells had damaged merchant vessels – likely to be the victims of both Russian and Ukrainian fire and possibly accidental hits. Mines sprinkled in the sea off Odesa reportedly sank at least one merchant ship.[7] The Russians also prepared to launch an amphibious assault against Odesa, assembling an invasion task group after declaring its local waters out of bounds to others.

A foretaste of aggressive action by Russian naval forces had been provided in late June 2021. The British destroyer HMS *Defender* – specially fitted with intelligence-gathering technology in addition to having the Sea Viper air-defence system – and which had just visited Odesa, dared to sail off the coast of the Crimea.

The Russians sent out strike jets and at least one flying boat to buzz *Defender*, claiming to have dropped bombs in the vicinity of the British warship. A Russian Navy corvette that had been monitoring the RN vessel for some time carried on shadowing at a distance while coast guard vessels got much closer.

According to Moscow, they even fired warning shots, though the UK denied that was the case. The UK also suggested no bombs were dropped near *Defender*, though there was a naval exercise being conducted by the Russians. During his annual radio phone-in with Russian citizens not long after the incident, Putin boasted that, had his military wanted to, they could have sent *Defender* to the bottom of the sea.

In the opening weeks of the Ukraine War, the risk of clashes between the navies of NATO and Russia was minimal, provided escalation ashore was avoided. Should Alliance and Russian air and land forces become embroiled in combat it would likely lead to conflict on the high seas too.

The Russian Navy had assembled a powerful task force drawn from its Northern, Baltic and Pacific fleets – including two large Slava Class guided-missile cruisers, sisters of the *Moskva* called RFS *Marshal Ustinov* and RFS *Varyag* – in the eastern Mediterranean. Its purpose was to counter three NATO Carrier Strike Groups (CSGs) led by the USS *Harry S. Truman* (USA), FS *Charles de Gaulle* (France) and ITS *Cavour* (Italy). Meanwhile, NATO was gathering together a powerful array of warships in waters off Norway for the long-planned Exercise Cold Response.

The British aircraft carrier HMS *Prince of Wales* was playing a key role in Cold Response, as flagship of NATO's Maritime High Readiness Force (MRF). The US Navy's USS *George H.W. Bush* CSG might also appear, replacing the *Harry S. Truman* CSG, which was meant to head for the Arctic too, but instead stayed in the Mediterranean. Ultimately, the Americans sent the assault carrier USS *Kearsarge* as a stand in for USS *Harry S. Truman* in the High North. Meanwhile, the Russians retained around 20 naval vessels in the Arctic, not least their best submarines but also including the nuclear-powered battlecruiser RFS *Pyotr Velikiy*. In the Black Sea, on 13 April 2022, the Ukrainians got revenge for the bombardment of Snake Island and their own navy suffering significant losses. They targeted the 12,500 tons, 40-year-old RFS Moskva with two anti-shipping missiles – allegedly deploying drones to distract the ship's radar operators – causing catastrophic damage to the Black Sea Fleet flagship and likely killing dozens of sailors.

A bungled bid to save the ship meant fires raged out of control. The survivors were evacuated to tugs standing by and *Moskva* went to the bottom on 14 April after an attempt to tow the mortally wounded cruiser back to Sevastopol failed.[8]

It was a heavy blow, the largest Russian Navy surface warship sunk since the Second World War and Moscow's most devastating naval casualty since the destruction of *Kursk* almost 22 years earlier.

The *Moskva*'s demise forced the Russians to place their warships well back from the enemy's land-based missile systems. More losses of people and equipment on and around

Snake Island were soon inflicted by the Ukrainians – using jets, drones and missiles – including several other Russian vessels eliminated.

After numerous strikes, on 30 June the Russians withdrew their forces from Snake Island altogether. They continued to use surface warships and submarines to bombard Ukrainian cities with long-range cruise missiles.

* * *

This was the situation in Europe and off its shores at the time of writing, with war raging in Ukraine along with heavily-armed vessels in close company and keeping a wary eye on each other from the Arctic to the Mediterranean.

The flame of hope for lasting peace between Russia and the West that had burned so bright in August 1991, when HMS *London* sailed to Murmansk and Archangel, had been snuffed out. With the West's nations looking to bolster their defences thanks to Putin's war in the Ukraine, it might well be the future *London* will have her armament and defensive suite enhanced. Construction of the next HMS *London* must also proceed with alacrity, as there is no time to lose in facing down the ravenous Russian Bear.

Notes

1. President Vladimir Putin quoted in the *Daily Mail*, 24 February 2022.
2. As reported in *WARSHIPS IFR*, October 2014. Putin was asked at a press conference what his reaction was to claims he had sent Russian troops into rebel areas of eastern Ukraine. He suggested that he did not need to answer for such actions, especially as his nation was 'strengthening our powers of nuclear restraint,' latter adding that 'it's better not to mess with Russia.'
3. As Putin described the invasion of Ukraine in his televised address to the Russian nation and also justifying it to the world. Taken from the official transcript of the 'Address by the President of the Russian Federation', 24 February 2022.
4. President Putin when addressing permanent members of Russia's Security Council, 3 March 2022. Putin had previously falsely labelled the Kiev government neo-Nazis and in this meeting suggested, without presenting any evidence, that Russian troops were fighting 'neo-Nazis who torture and brutally murder prisoners.'
5. *iNews* – 'What is Russia's nuclear deterrent? How many weapons Putin has and what putting forces on high alert means' – online news report: https://inews.co.uk/news/world/russia-nuclear-deterrent-what-putin-how-many-weapons-ukraine-high-alert-explained-1488254
6. 'Address by the President of the Russian Federation', 24 February 2022.
7. *Naval News* – 'Estonian Cargo Ship Sinks Off the Coast of Odessa,' online news report: https://www.navalnews.com/naval-news/2022/03/estonian-cargo-ship-sinks-off-the-coast-of-odessa/
8. '*Moskva* Loss was the Worst for the Russian Navy Since *Kursk*', a report and analysis of RFS *Moskva*'s sinking, *WARSHIPS IFR*, June 2022.

APPENDIX 1
LONG VOYAGE TO AN EPIC FIGHT AT WAR'S END

Following publication of the hardback edition of this book in 2003, two men who served in the cruiser HMS London *during the Second World War and the son of another, who had been her ship's doctor during the Yangtze Incident, got in touch with me. They passed on new information including their own recollections of key events in the ship's history.*

We start here with the story of a young officer who ended up involved in a very famous sea fight, then learn how the ceremonial sword destined to be presented to Stalin (see p.123/124) got 'lost'. Finally, there is a boy's unusual mode of transportation to welcoming his dad home from the post-war deployment that saw London *attacked by communist Chinese forces (see Chapter Sixteen).*

Long Voyage to an epic Fight at War's End
Captain John Beckett Robathan loaned me his midshipman's journal – sending it in the normal Royal Mail post, which was quite an

Captain John Robathan during the 1970s.
Photo: Private Collection.

Midshipman John Robathan is pictured back row, second from the right in this hockey team photograph, taken aboard the cruiser (in front of the ship's battle honours board) in 1944 or 1945. Possibly with the ship's Commanding Officer in the centre, front row. *Photo: Robathan Collection.*

alarming development – with permission to use material from it. Sadly, shortly afterwards Capt Robathan crossed the bar, so we never got to meet in person but I am pleased to be able to tell the story of his time in the cruiser *London* here.

Like all midshipmen's journals of the time, Robathan's included not only diary entries, giving his impression of life aboard ship along with a perspective on events both interior and exterior to the vessel, but also sketches of warships, harbours and other things along with photographs and documents.

Overall Robathan's journal provides an amazing ringside seat for some of the key operations by the Eastern Fleet of the Royal Navy – a Cinderella force since 1942, but which in 1944 was being revived to take the war to the enemy.

* * *

Having entered the Royal Navy in early 1943, January 1944 saw eighteen-year-old Midshipman Robathan drafted to the famous cruiser HMS *Belfast*, which the previous month had helped hunt down and destroy the German battlecruiser *Scharnhorst*. Robathan was almost immediately sent south from Scapa Flow along with other midshipmen, to join HMS *London* at Rosyth.

London was in dockyard hands, part of a great preparing of various British vessels for service in the Far East as the need for so many RN ships in European waters declined with the Nazi surface raider threat fading.

By mid-February HMS *London* was at Greenock and soon left for the long voyage to the Far East. Calling at Gibraltar – the British naval base bastion guarding the entrance the Mediterranean – the cruiser departed from there on 23 February and headed to Algiers where she took aboard 280 passengers for passage to Malta. It was when the *London* reached at Alexandria that Robathan received his first sight of a warship recently damaged in battle, writing in his journal:

The heavy cruiser HMS *London* at Scapa Flow, early 1944. *Photo: Robathan Collection.*

The wretched [cruiser] Birmingham was there still with an untouched hole in her bows, as result of being torpedoed [by U-407] early last December. She has not paid off, and the crew must be browned off with lying in harbour these last few months.

With this timely illustration of the purgatory a torpedo hit could inflict, HMS *London* zig-zagged – to hopefully make herself less of an easy target for any lurking enemy submarine – as she headed for Port Said. Beginning a transit of the Suez Canal on 1 March, Robathan found it to be 'much wider than I had imagined' with the ship steaming at 'about 12 knots, passing a few tankers going in the other direction.'

By that afternoon *London* was at anchor in the Great Bitter Lake, waiting for some merchant vessels going north to pass by, not far from two of Italy's Littorio Class battleships (*Italia* and *Vittorio Veneto*), which were now hoping to fight alongside the Allies. The Italian government was desperate to see them involved in the war against Japan, to at least garner some laurels before the conflict ended. It was an ambition that would not be realised, though the one-time Vichy French battleship *Richlieu* would come out to join the effort.

In the meantime, *London* reached Suez just as night fell on 1 March, heading out into the Red Sea. On 3 March Captain Richard Symonds-Tayler issued the *clear lower deck order*, assembling all but the engine room watch-keepers on the upper deck. Capt Symonds-Tayler told the crew their destination was Colombo in Ceylon, in company with the heavy cruiser HMS *Cumberland*.

However, *London* soon had to turn around and return to Suez at 28 knots to land a Royal Marine with appendicitis who needed an operation or he might die.

After that and calling at Aden, *London* headed out into the Indian Ocean, with some anti-aircraft gunnery practice on the way, reaching Colombo on 13 March.

The following day there was a visit by the Naval Commander-in-Chief of the new South East Asia Command (and also Eastern Fleet boss), Admiral Sir James Somerville. He appeared impressed with the cruiser's outfit of high-angle anti-aircraft weapons, for it was Japan's air forces the Allied navies had learned to fear the most.

When he addressed the ship's ratings and junior officers, Somerville – whose son John had, coincidentally, been a midshipman in the *London* in the late 1930s – proposed that the Japanese had 'missed the bus, as regards beating us up on the sea,' or so recorded Midshipman Robathan. The Eastern Fleet was now 'more than just a name' according to Somerville who had commanded it since March 1942.

Back then the Eastern Fleet's strength was so weak, and its battleships of the Royal Sovereign Class of such dubious worth, that it had to hide from the Japanese when, in April 1942, they rampaged across the Indian Ocean. They sank numerous Allied warships including the British cruisers *Dorsetshire* and *Cornwall* along with the carrier *Hermes*. Attempts by Somerville's fleet to hit back had been ineffective, but by early 1944 – with the tide turning against the Axis powers – it was being rebuilt and looking to become aggressive.

As part of this new, more assertive behaviour the Eastern Fleet, including *London*, deployed to sea for Operation Diplomat on 21 March, which was expected to last around ten days. It was a rehearsal for combat with the enemy, including how to launch large strikes by carrier aircraft and the skill of refuelling from Royal Fleet Auxiliary (RFA) tankers. It was also a means of exerting a deterrent presence close to routes used by Allied merchant shipping.

A drawing of the Eastern Fleet battlecruiser HMS *Renown*, by Midshipman John Robathan for his journal. Robathan spent time in the *Renown*, making a voyage in her to transfer back to *London* in spring 1944. *Image: Robathan Collection.*

'The point of the operation was to guard our trade link between Australia and India and also to be in the offing,' according to Robathan in his journal.

It was hoped the Japanese might be provoked to come out and the Eastern Fleet would get its chance to deliver some pay-back for 1942. It would also be a baptism of fire. The crews of the major warships, which had all come out directly from the UK, were for the most part inexperienced and had never seen combat.

Battlecruiser *Renown* was flagship, with Vice Admiral Sir Arthur Power, second-in-command of the Eastern Fleet, sailing aboard her. The task force's other heavy firepower was provided by the battleships *Queen Elizabeth* and *Valiant* while air cover came from the carrier *Illustrious*. Cruisers *London*, *Gambia*, *Ceylon* and *Cumberland* along with escorting destroyers rode shotgun. *London* was usually stationed astern or to starboard of *Queen Elizabeth* and on 22 March there were 'one or two submarine contact scares' which provoked 'much manoeuvring', according to Robathan. Otherwise the training went on as planned, *London* and other ships training in air defence, with aircraft from the carrier pretending to be Japanese attackers.

The Dutch cruiser HNLMS *Tromp* and her escorting destroyer HNLMS *Van Galen* joined the task force, in company with three British fleet oilers and on 25 March *London* and other ships practiced replenishment-at-sea, an exacting manoeuvre with both ships linked by hoses. The latter was a necessity for any fleet operating across the vast expanse of the Indian Ocean or Pacific, thousands of miles from the nearest home base. Disturbing the smooth running of the various exercises at 5.35pm was another submarine alarm, as reported by Midshipman Robathan in his journal.

…an echo was picked up by the Asdic operator, and it was supplemented with electric motor noise and doppler effects. We altered course to pass over the submarine and depth charges were brought to the ready, but the Captain said they were not to be dropped, so nothing was done, not even a report [made]. *Odd.*

On 27 March the Eastern Fleet received an eagerly anticipated reinforcement when the US Navy carrier USS *Saratoga*, a veteran of several battles, made a mid-ocean rendezvous, also bringing the destroyer escorts USS *Dunlap*, USS *Fanning* and USS *Cummings*. The *Saratoga* had been loaned to Somerville's fleet not only to boost its strike power, but also to pass on the USN's vast experience in mounting complex air operations from the sea in the face of the enemy. The Royal Navy had much to learn in that respect and the Americans were more than happy to 'help the British initiate their carrier offensive in the Far East.'[1]

More exercises followed over the next few days and on 29 March *Illustrious* took the lead, launching and recovering many planes, which also provided several more opportunities to practice anti-aircraft procedures.

By 30 March Operation Diplomat was coming to an end, with expectations of an enemy response, but, as Robathan noted 'the Japs had not come out and consequently nothing of interest happened.'

Yet even though the enemy was not tempted to try his luck, Robathan thought 'the operation helped to knot the EF [Eastern Fleet] together and give everybody a taste of what it is like to operate in a fleet. Besides, the USS *Saratoga* was initiated into the EF, and she is bound to be a most useful addition…'

The Eastern Fleet ships headed for Trincomalee and waiting for them there was the cruiser *Nigeria*, which Robathan noted had 'come straight out from Scapa in our footsteps' while big gun firepower was boosted by the arrival of the French battlewagon *Richlieu* on 12 April.

It seemed on 15 April that finally there might be some action on the horizon, with Robathan writing in his journal that a 'buzz was started, when the C in C [Somerville] suddenly took his abode in the *Queen Elizabeth*; everybody felt something was about to happen. The Americans ashore in the evening were saying how we were going out on the 'morrow to beat up Sumatra, which turned out to be correct…' Robathan added: '*Saratoga* told its chief petty officers at noon on Saturday that they were going to attack an island off Sumatra: they appear not to be able to hold their tongues if that is the case.'

This was Operation Cockpit, which was scheduled to be unleashed on 19 April, against the port of Sabang, on We Island, to the north-west of Japanese occupied Sumatra. The aim was to damage the port infrastructure and also destroy enemy oil reserves. 'The fleet was to lie off 100 miles, whilst aircraft from the carriers carried out their attacks,' explained Robathan.

The fleet took its time in getting there, giving *Saratoga*'s aviators a chance to work closely with their British counterparts 'to impart some of their experience to the British pilots.'[2]

On D-2 (17 April), according to Robathan, the cruisers and battleships 'oiled the destroyers, making quite sure they would be topped up enough , so that they could get back without any help,' with *London* providing fuel oil to HMS *Quadrant*.

On D-1 (18 April), which happened to be Robathan's nineteenth birthday, he related that 'everyone was getting prepared for action; morphia, action rations etc issued; mess decks cleared of all clothing etc.' On D-Day (19 April) the call to hands came at 4.30am, courtesy of Royal Marine buglers over the public address system, with the ship's company going to action stations at 5.00am.

Signals interception efforts had picked up an enemy submarine, which seemed to be in the Eastern Fleet's vicinity, sending a message, which was possibly a sighting report, to an enemy headquarters

Admiral Somerville had earlier sent a general message to the Eastern Fleet saying that he believed they had not been sighted and, so Robathan recorded, according to the admiral 'with luck, we would catch Japs with their kimonos up.'

The air strike was launched at 5.30am, with 15 Barracudas from *Illustrious* and 30 Dauntless and Avenger bombers from *Saratoga*, escorted by Hellcat and Corsair fighters.

Admiral Sir James Sommerville (centre) chats with the Commanding Officer of USS *Saratoga*, Captain John H. Cassady, during a visit to the American carrier at Trincomalee in April 1944.
Photo: US National Archives (80-G-K-15604).

The launching of this group was followed 15 minutes later by a dozen Hellcats assigned to ground-strafing. Over the fleet were Corsair fighters on Combat Air Patrol (CAP) in case of enemy air attack. By 7.00am the strike aircraft were returning with no losses reported, though, according to Robathan's journal, one Hellcat had actually been shot down but with the pilot picked up by an Allied submarine. In the fleet there was no relaxation, with Robathan noting 'we settled down to wait for the expected retaliation.'

Just after 10.00am unidentified aircraft were detected by radar circling at a distance and Hellcats launched to go and investigate, shooting down two Kate torpedo-bombers of the Japanese naval air arm.

'A few minutes later another Kate 10 miles astern was faithfully dealt with by 4 Hellcats,' reported Robathan, 'and a shout went up from all those on the upper deck when they saw it crash, making a large column of smoke. I hoped they had not shouted too soon: it might have easily been an American, but luckily it wasn't.'

And so they waited, with those closed up below decks in *London* enduring stifling conditions, especially in the shell handling rooms which were like sweatboxes. Men were allowed up onto the upper deck in relays for half an hour in the fresh air.

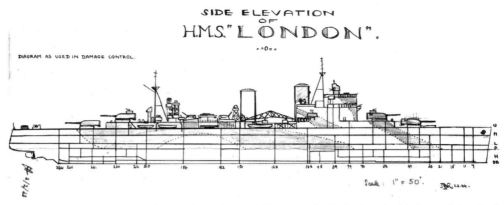

A side elevation of HMS *London*, as drawn by Midshipman John Robathan for his journal, inspired by a damage control diagram displayed in the warship. *Image: Robathan Collection.*

Two more mystery aircraft were detected but it transpired they were Allied. When darkness fell the enemy made his most ambitious forays against the fleet, which unfolded as follows according to Robathan:

> *...at 19.33, we opened a barrage fire with 4", 8" and pom-pom* [guns] *against a few enemy aircraft which were flying low down quite close. The fleet put up quite a tidy barrage, although it did not look very dense...it was all over by 20.15...the only danger seemed to be that of firing into your own ships.* Gambia *nearly took a packet from one of our barrages.*

On D+ 1 (20 April) Robathan noted 'there was an aircraft scare at 01.20' but the enemy scout – some 25 miles away – failed to spot the fleet. Danger over, Somerville sent a signal to the Admiralty that informed their Lordships the fleet had given 'Sabang a damn good bang.'[3]

*   *   *

The Eastern Fleet returned to Trincomalee, where Robathan suffered heatstroke while tackling an assault course ashore with his fellow midshipmen.

He ended in in a naval hospital ship for a few days. While he was there, he and another midshipman from the cruiser (also receiving treatment for heatstroke) heard that *London* had gone to Australia for two months.

A disgusted Robathan remarked in his journal:

> *Very funny, I don't think! We are not in the slightest bit amused.*

There might be a chance of a flight to Australia in a Catalina flying boat, but that was only really for VIPs. Robathan and the other midshipman ended up going to *Renown* on 6 May 1944, with the chance of returning to their own ship once the fleet assembled for an operation. In the meantime, Robathan was surprised by the informality of the dress code in the battlecruiser. 'For a flag-ship, it is not at all "pusser". Officers wear khaki with or without shirts and usually very scant sandals: on watch too.' *Renown* was poorly ventilated, which was not an uncommon feature in warships constructed a long time ago for fighting the Germans in the North Sea. The *Renown* sailed, leading Force 65 and heading for a rendezvous with Force

66 – the rest of the Eastern Fleet – to mount Operation Transom, a raid on Surabaya, which aimed to strike oil refineries and also an important enemy naval base.

The *Renown* and her retinue headed for Exmouth Gulf in northern Australia – which had become a refuelling station for the British ships – where Robathan and the other midshipman were transferred to HMS *London*.

Along with the cruiser *Suffolk*, the *London* had earlier been tasked with escorting half a dozen fleet tankers to Exmouth Gulf, so they could top up in preparation for keeping the Eastern Fleet at sea in coming weeks.

D-Day for the strike was 17 May and after preparing for action the previous evening, Midshipman Robathan turned in.

After a pleasant night's sleep on deck I awoke at 03.30 at the reveille. We went to action stations at 04.00; at the first signs of light, aircraft were flown off. The blue flame from their exhausts was very noticeable. The strikes were to consist of 18 Avengers, carrying bombs, and a dozen Corsairs, from Illustrious; *24 Hellcats, 12 Dauntless and 18 Avengers from* Saratoga. *The total weight of bombs would be just about 30 tons. It took some time for this lot to fly off, assemble and gain height and it was 07.20 before the last strike was ready to set out on its 170-mile trip to the northward and Sourabaya [sic].*

From his action station on the upper deck, Robathan watched as an Avenger dropped from the formation 'with a thin wisp of smoke trailing behind it.'

It came down, the smoke stopped and did two circuits of Illustrious, *the engine running before making a neat crash landing on the water on* Illustrious' *starboard side ahead of us. It made a pretty hefty splash and it seemed pretty miraculous, when we saw the crew of three emerge in a rubber dinghy, the plane having sank, bombs and all, in about 10 seconds. They passed down our starboard side, marked by a smoke float dropped by the carrier…they were soon safely picked up by the destroyer HMAS* Queensborough.

The fleet waited for the strike force to come back protected by an 'air umbrella of 8 Hellcats and 8 Corsairs.' At 9.15am, aircraft were picked up on radar but identified as friendly. It was the strike force beginning to return and by 10.00am aircraft were landing back aboard the carriers. By 10.30am, the fleet had set a course for Australia, noted Robathan, 'and everyone was just waiting for what was coming to us [from the enemy]. But it never came.'

On 18 May farewells were said to USS *Saratoga*, which, after a spot of R&R, was to head back to the USA, for a refit at Bremerton. Hundreds of sailors and Royal Marines assembled on *London*'s upper deck to cheer the American carrier on her way.

USS *Saratoga* with her three escorting destroyers steamed down the port side of the British cruiser and other ships in the Eastern Fleet. *London* sent a message by light signal: '*Saratoga*, Sabang, Sourabaya, Swell. Good luck!'

Thousands of sailors and aviators crammed onto *Saratoga*'s flight-deck roared their lungs out in return, the ship's band playing 'God Save the King!' as a salute to their Royal Navy shipmates. 'By dark, the vast smokestack of the *Saratoga* had disappeared below the horizon,' noted Robathan in his journal, 'as she steamed to the south eastward to Fremantle for three days recreational leave.' The British were heading for Exmouth Gulf and, after refuelling, back to Ceylon.

* * *

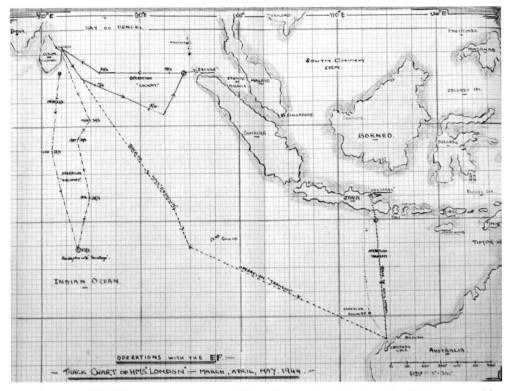

A track chart of HMS *London*'s operations with the Eastern Fleet, March – May 1944, as created by Midshipman John Robathan for his journal while serving in the ship. *Robathan Collection.*

With HMS *London* operating out of Colombo on humdrum training and convoy escort duties, Robathan used his journal to recorded on 6 June the fall of Rome and the Allied invasion of northwest France. The Italian capital had been declared an open city, with German forces withdrawing. Robathan noted that the Pope was cheered for that great happening 'as if he did anything!' He was more pleased to write that 'the famous Second Front has started at last' in Normandy.

Great events were certainly unfolding on the other side of the world, but aside from 48 hours leave in Fremantle, the weeks aboard *London* in the Indian Ocean 'passed uneventfully'. Then, disaster befell one of the British battleships.

In early August 1944 a floating drydock at Trincomalee collapsed with the *Valiant* inside it, almost taking the battleship down too as it fell apart.

HMS *London* was at sea when it happened, returning on 9 August with Robathan among those who gazed at '*Valiant* down by the bows' swinging at a buoy 'with divers down all the way along her starboard side. Of the great floating dock, nothing could be seen, but the tops of cranes sticking out of the water' which, so Robathan recorded, was covered in debris and oil.

To further broaden Robathan's naval education, on 12 March 1945 he was drafted to the destroyer HMS *Venus* in which he was to serve for some months. It was aboard her that, on 9 May 1945, Robathan listened to King George VI make a speech on the end of the war, thinking of home and wishing he was there 'to join in the celebrations and hear the church bells' peal out for Victory in Europe.[4]

For Robathan and other men in the Eastern Fleet the war was to drag on for months while people back home began returning to peace time ways, though for him its most dangerous and exciting moment was imminent.

Having been just an onlooker so far during the war – seeing major events and skirmishes from a distance – Midshipman Robathan was soon to be plunged into combat in the most spectacular style. HMS *Venus* was part of the 26th Destroyer Flotilla, led by HMS *Saumarez*, and also including HMS *Verulam*, HMS *Vigilant* and HMS *Virago*. To them went the honour of engaging enemy surface warships in the Royal Navy's last full-blooded destroyer action.

The heavy cruiser IJNS *Haguro* had been spotted by British submarines heading for a besieged outpost on Andaman Islands where she was to try and evacuate Japanese troops. A hidden hand was also at work. Spearheaded by the Royal Navy's Y Service, whose Far East-based female sailors were highly adept at their job of enemy wireless signal intercepts and code-breaking, secret intelligence was guiding the British bid to trap and destroy *Haguro*.

The Japanese cruiser IJNS *Haguro*, under attack from American aircraft at Simpson Harbour, Rabaul, late 1943. *Haguro* may have escaped destruction then but in May 1945 was sunk by a flotilla of British destroyers including HMS *Venus*, the latter with John Robathan (who had just transferred from HMS *London*) among her crew.

Photo: US Air Force, from the collections of the US Naval History and Heritage Command/NH 95558.

After finishing his time in HMS *London*, Midshipman John Robathan served in the destroyer HMS *Venus* when she helped to sink the Japanese cruiser IJNS *Haguro* during the Battle of the Malacca Strait. *Photo: Robathan Collection.*

On the night of 15/16 May 1945 the 26th Destroyer Flotilla, whose own signals intelligence specialists had been monitoring enemy radio transmissions and radar operators picking up contacts, went up against the enemy heavy cruiser in the Malacca Strait.

A classic example of how a well co-ordinated force of smaller destroyers could kill a big warship, it would be used as a text book example during tactical lessons at naval colleges for decades.

The pack of British destroyers surrounded the Japanese cruiser, the fight taking place during ferocious lightning and rain squalls. There were multiple torpedo launches amid the thunder and flash of guns adding to nature's bedlam. At one point, Robathan saw 'a huge mushroom of red flame billowing up into black smoke' erupting from the *Haguro*, 'followed by dying down to a red glow.'[5]

When the gun he was assigned to jammed, Robathan went up to the bridge where the captain told him to stand by a voice-pipe ready to pass on orders.

With two torpedoes left, *Venus* lunged at *Haguro*, all guns blazing as she steamed in. The rain having ceased, but with clouds pressing down, there was still a good view of the dying enemy warship – already stopped and sinking, with water washing over the upper deck and black smoke pouring out of many jagged holes.

With star shells bursting overhead, fired by *Venus* to ensure aimers could see the target with as much clarity as possible, the destroyer launched her torpedoes.

Robathan watched for 'a tense minute, which dragged on for hours, before one after the other, two huge grey shapes leapt into the air, looking for all the world like poplars, one hit the stern and one hit the bow.'[6]

With *Venus* having delivered the *coup de grace*, the British destroyers – which had suffered just one of their number damaged (*Saumarez*, with three men killed) – saw the sea finally swallow up the *Haguro* just after 2.00am.

Low on fuel, the British departed the scene without picking up any Japanese survivors. The enemy destroyer *Kamikaze*, which had earlier been escorting *Haguro* but was damaged and ordered to withdraw, returned to save 400. Around 800 Japanese went down with the cruiser.

* * *

After the war John Robathan became a naval aviator and was transferred to the Royal Australian Navy, by 1948 finding himself at the controls of a Sea Fury fighter-bomber operating from the carrier HMAS *Sydney*.

In the early 1950s Robathan returned to the RN, continuing his career in the air and at sea, flying the Sea Venom fighter jet as a squadron commander.

Robathan made it his habit to come back aboard the carrier last after seeing the rest of his squadron home safely. He told one Observer who flew with him, and who was worried about their aircraft running out of fuel (and having to ditch in the sea), that he should just keep an eye on the rev counter. Robathan added: 'The moment you see it start to wind down [indicating that the engine was stopping for lack of fuel], bang out [eject]. I shall be a split second behind you.'[7]

As his time as a naval aviator came to an end, while flying from a carrier based at Singapore, Robathan took his jet up to 42,000ft, putting the aircraft into a vertical dive and, as noted in his logbook, 'went supersonic March 1.3.'[8]

After a spell as captain of the frigate *Jaguar* (1963/64) Robathan was appointed Captain of the Air Department in the carrier *Victorious* (1966/67) then commanded the destroyer *Undaunted* (1970/71).

Robathan, with typical dry humour, referred to her as 'the Undented.'

In the same ship, and while he was Captain (D) – the senior destroyer captain – of the Portland Squadron, in 1971 Robathan took its vessels into the Baltic for a visit to Copenhagen.[9]

It was while serving as the Captain of Royal Naval Air Station (RNAS) Culdrose in Cornwall (1971-73) that Robathan came up with an innovative solution to avoid being late for a meeting ashore. At sea yachting – one of his favourite pursuits – Robathan found his boat becalmed by light airs. 'His solution was to call Culdrose and order an air sea rescue exercise and, as a helicopter hovered over his yacht, he jumped into the sea.'[10] He was winched up and flown to the air station.

Next came command of the County Class destroyer *Kent* (1973-75), which was at one time dogged by a shadowing Russian destroyer in the Mediterranean. Seeking to take the tension out of the situation, Capt Robathan ordered his embarked helicopter to pay the vessel a visit.

The helicopter hovered above the shadower, using its winch to lower a bottle of whisky sent across from Robathan to the cheer the Soviet vessel's CO. Having hoped for a bottle of vodka in return, Robathan was disappointed that his Russian opposite number did not see the joke.

On coming ashore for good, Robathan was appointed an Aide to Her Majesty the Queen. Later, in retirement, Capt Robathan served on as naval representative in Cornwall and did not finally hang up his blue suit until 1993.

Capt Robathan passed away in 2005, but I was able to return his journal to his widow, Anna, in person during a day trip to Cornwall. It was a great relief to hand over custody of such an important historical document, but I am so glad it gave me an opportunity to include Capt Robathan's adventures here.

A Young Officer's Sword of Damocles

Lieutenant Colonel T. P. Furlonge served in the cruiser *London* as her Captain of Marines from April 1943 to October 1945.

'I thought you might be interested in some of my recollections, which I have never [until now] recorded,' he wrote to me in March 2013.

Lt Col Furlonge explained that his previous ship had been the old carrier HMS *Furious*, which was crewed by a mix of men from the West Country and Liverpool. They seemed a lot more laid back than the 'volatile' men of the *London*. The cruiser's sailors and Royal Marines never spoke about their ill-fated summer of 1942 experiences in the Arctic, but, so Lt Col Furlonge admitted, this 'reticence was not particularly noticeable to me at the time'. He added that it was 'not surprising as I was only 22-years-old and not particularly sensitive. I only discovered that *London* had participated in [convoy] PQ17 recently.'

One vivid memory for Lt Col Furlonge related to the so-called 'Stalingrad Sword', which he recalled caused him 'both anxiety and embarrassment.'

> *On the trip out to Alexandria we took with us a number of female passengers, some of whom [from the Women's Royal Naval Service, the WRNS] manned the cipher office. One afternoon, shortly after departure, the cipher officer called me at my action station which was halfway up the foremast. He needed the cipher book urgently. I told him to go to the cipher office and look for the keys on the fourth hook on the shelf above the typewriters, go to the store, and collect the book. He called a few minutes later saying that he could not find the keys and at this point I realised that I was facing a crisis.*

This was because, awed by the significance of the Stalingrad Sword, which Winston Churchill was intending to present to Stalin at Tehran, the young Royal Marine officer had put it somewhere he felt was totally safe.

> *I had stowed the Sword in the Confidential Bookstore which I considered to be the most secure place in the ship. Failure to produce it when we reached Alexandria was unthinkable and I appreciated that the situation was advancing towards a diplomatic disaster. The only method of entering the store without unlocking it was by using an oxy acetylene cutter, but as I could not remember exactly where I had put the Sword, the possibility that Churchill might have to present the Sword in a slightly singed case crossed my mind. A frantic search of my cabin and anywhere else I might have put the keys took place for the next 24 hours. The Captain was aware and anxious. Eventually I undertook a more thorough search of the Cipher Office and found that, instead of being hung on the fourth hook from the left, they had been hidden away at the back of the stationery drawer we never used.*

Furlonge's suspicion was aroused by the tidy-minded temporary inhabitants of the cipher office. 'My verbal expression of disapproval of having aboard women who tidy things up was the subject of a formal complaint by one of the WRNS officers.'

Lt Col Furlonge also related how he noticed something lacking in his own behaviour in the image on p.130 of this book. It shows the Royal Marine honour guard arrayed on *London*'s quarterdeck as enemy senior officer arrives to negotiate surrender terms at Sabang.

Lt Col Furlonge noticed that his younger self was 'standing at attention and not saluting the Japanese Admiral when he came aboard. Your book has therefore provided evidence of my insubordination.'

Lt Col Furlonge's verdict on his time in the cruiser was as follows:

'*London* was never as hectic or exciting as *Furious*, but I survived. We were cold, uncomfortable and frequently terrified but not everybody had the opportunity to be heroes.'

Yangtze Incident Doctor

Ian Taylor wrote to me with a correction to his father Surgeon Commander Wilfred Taylor's name – taken from his 'London Gazette' award citation in the hardback edition (*p.153*), but amended in this paperback to reflect that he liked to be called W.B. 'Bill' Taylor. He was an eye specialist by trade, seeing service during the Second World War in South Africa, but by early 1945 assigned to Royal Naval Hospital (RNH) Lowestoft to look after the sailors of the RN Patrol Service, then moved down to RNH Stonehouse in Plymouth.

After promotion to Surgeon Commander in 1947, Taylor was the following year appointed Principal Medical Officer in HMS *London*.

While it is the account of Bandmaster Harwood taking care of the wounded that features heavily in my account of the Yangtze Incident (and *London*'s torrid time under Chinese fire) Surgeon Commander Taylor, was obviously in the thick of it too. Regretfully, according to Ian, his Medical Log is not currently accessible.

However, Ian Taylor was able to observe of his father's part in the Yangtze Incident, that 'in common with Bandmaster Harwood, he went without sleep to operate on those casualties during the engagement. Our family hearsay is 72 hours.' Within his father's Medical Log, Ian Taylor recalls, 'if I remember correctly, there were nearly fifty casualties listed.'

But the most vivid memory for Ian Taylor himself comes from the homecoming of the cruiser after the Yangtze Incident. 'I remember as a very small boy of five, being passed over the heads of the crowd to reach the gangway to be able to greet my father when families met the ship at Chatham.'

Notes

1. Dictionary of American Naval Fighting Ships (DANFS), USS *Saratoga*, Ship History, US Naval History and Heritage Command (NHHC). https://www.history.navy.mil/research/histories/ship-histories/danfs/s/saratoga-v.html
2. Ibid.
3. Donald Macintyre, *Fighting Admiral*.
4. Robathan's private diary, as quoted in *Sink the Haguro* by John Winton.
5. Ibid.
6. Ibid.
7. Capt J. B. Robathan, unpublished obituary, courtesy of Peter Hore.
8. Ibid.
9. 'Navy News', August 1971.
10. Capt J. B. Robathan, unpublished obituary, courtesy of Peter Hore.

Command HMS *London* 'in action'

These days the Second World War heavy cruiser lives on in the Russian-origin *World of Warships* computer game where, should you wish to command the cruiser in simulated combat, you can jump aboard. Visit: https://worldofwarships.eu/en/news/history/armada-london/

It's mass market interactive entertainment but the ship is brought to life in stunning fashion by *World of Warships*.

There is even a World of Warships video on HMS London: https://www.youtube.com/watch?v=KD-A0EAKq5w

Worth a look too

For a close up look at cruiser *London* via Second World War footage, it is worth watching this video posted by Armoured Carriers on YouTube https://www.youtube.com/watch?v=gJIp7kQ8D0o

British Pathé have some interesting clips of the cruiser too, from launch in 1927, through the Second World War to her return from the Yangtze Incident in 1949, plus also the guided-missile destroyer of the 1960s. Visit: https://www.britishpathe.com

APPENDIX 2
A *LONDON* RISES AGAIN FROM THE MURKY VOID

Richard Endsor's side elevation of the *London* (1656), as reconstructed from architectural drawings of the ship in the Royal Collection. *Courtesy of Richard Endsor on behalf of The* London *Shipwreck Trust.*

The story of the wooden wall warship named *London* that saw service in both Oliver Cromwell and King Charles II's navy, and which blew apart in 1665, is not yet over.

There have been a few more twists and turns, including one in recent years that threatened a second catastrophic explosion. We will return to that extraordinary episode later, after recording how in recent times this particular *London* has declared 'resurgam!'

*   *   *

In 2005, two years after publication of this book's hardback edition, the remains of the *London* that was built at Chatham and launched in 1656, only to be ripped asunder nine years later, were explored by divers.

Lying a few feet under the surface off Southend-on-Sea, not far from the Essex seaside resort's famous one-and-a-quarter-mile-long pier, the scale of treasures the ship could yield were finally understood.

For, despite the passing of 340 years at the mercy of the sea, it appeared what still lay in the wreck could yet rival the finds of the famous *Mary Rose* at Portsmouth and the equally significant *Vasa* in Sweden.

'the ship exploded in searing heat and flame'
At the time of *London*'s demise in 1665, a recent major refit had just concluded (*see P3 of this book*) and guns were being loaded aboard for a hotly anticipated fight with the Dutch.

It is worth elaborating further here on how it is likely the ship met her end. According to the historian Richard Endsor, author of several books that bring the Restoration Navy of 17th Century England vividly to life, disaster fell on the *London* when her people were least expecting it.

'Near land and out of harm's way in the river, life aboard would have been relaxed and noisy, no doubt helped along by the generous allowance of beer,' Endsor has written.[1]

> Many men would have been stowing the stores below decks. Some men would have played music on their own instruments, probably to the tune of popular songs while others played board games. In quieter corners, many women – for they were allowed aboard during the fitting out – would have been saying their goodbyes to husbands and lovers, hoping for their safe return.

As related earlier in this book, the ship and most of her people were snuffed out in an instant, a cataclysm described by Endsor in vivid terms:

> Suddenly, the ship exploded in searing heat and flame, within minutes ending up on the bottom in two distinct parts. Death would have been mercifully sudden for most. The dreadful catastrophe must have been caused by an accident involving the tons of gunpowder stored below decks in the magazine.

Remains of those who died in the *London*'s catastrophic explosion are reckoned to still lie within the wreck and in 2008 it was safeguarded under provisions of the Protection of Wrecks Act 1973.[2]

As noted by Endsor, a number of *London*'s 'valuable guns' were actually salvaged 'almost immediately' after she was destroyed.[3] That process was carried on into the 1680s, though it was not until 1820 that the wreck's site was marked with a buoy. From the early 1960s it was more than once surveyed, which is when it was confirmed that *London*'s remains were in two sections, about 400 metres apart.

A degree of instability was recorded and there were concerns the wreck may shift and foul passing ships. Heavy traffic in an adjacent shipping lane meant major disturbance. Then there were the actions of dredgers and strong currents sweeping away history, with many precious items likely lost due to the wreck breaking up. That was why declaring it a protected site was urgent.

The case was made for *London* being of great national and global historical importance. This succeeded and a decision was taken to divert the Thames shipping channel further away to try and reduce further damage, to at least slow the pace of erosion and prevent more artefacts disappearing into the murky void.

In 2015 – 2016, thanks to collaborative work by Historic England, Cotswold Archaeology, 'Licensee divers' and Southend Museum along with volunteers, 700 items were brought to the surface.

The *London* Shipwreck Trust was established in 2016, by experienced civilian divers specialising in maritime archaeology. They had been diving on the wreck since 2011, with a mission statement of supporting 'the crucial recovery of artefacts and remains of the *London* before they are lost forever.'

* * *

Southend Museum is the official receiving museum for artefacts, while Historic England is tasked with managing the wreck. It has authorised recovery of artefacts, on the understanding that the Save the *London* 1665 group handles funding their conservation. As a result a fund-raising campaign was launched by the Nautical Archaeology Society on 3 July 2019 – the 363rd anniversary of *London* herself being launched by the Admiralty.

Due to being covered by silt, wooden objects and leather items that would never have been so well preserved on land, have stayed in a remarkable condition.

Historic England has revealed they include 'pewter spoons…and a urethral syringe' along with other objects such as Gunners' equipment (including a handspike). A flexible rammer, small arms and a gun carriage truck, made of wood and brass have also been recovered.[4]

The major importance of relatively small objects has been stressed by archaeologist Phil Harding. A familiar face to television viewers, thanks to his work on Channel 4's 'Time Team', he has been made Ambassador for the Save The *London* 1665 campaign in addition to being a Trustee of The *London* Shipwreck trust.

Harding has explained that 'one of the most important things about the *London* is not necessarily the vessel itself, but the objects that we can recover... They reflect life in 17th Century England.' Harding added: 'The beauty of it is that what we are really dealing with is not just a ship and a piece of archaeology…we're talking about British history as well. The sheer fact that it exists and, more importantly, that it's so well-preserved means that this is an absolutely vital piece of our maritime heritage.'[5]

London's site is not an easy one to dive on at the best of times, according to Steve Ellis of the *London* Shipwreck Trust.

Most people think we are actually bonkers diving the Thames…It's mostly really fierce tidal currents and the visibility is mostly pitch black. But, the thing is, sometimes you get a really special moment…you just do not know what you're going to come across and to have this magnificent ship for me to dive on, as an amateur archaeologist, is a dream come true.[6]

Intrepid diver-archaeologists Steve Ellis (right) and Steve Meddle, with a gun carriage wheel recovered from the wreck of the *London*. *Photo: Courtesy of Steve Ellis.*

A sun dial pockets compass recovered from the wreck of the *London* in the Thames. *Photo: Courtesy of Steve Ellis.*

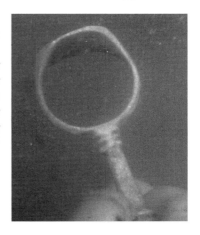

The moment of discovery in the murky waters of the Thames as Carol Ellis finds a signet ring pipe tamper used by one of the *London*'s crew. It was preserved in the silt of the Thames for more than three centuries. *Photo: Courtesy Steve Ellis.*

Steve's wife Carol has also dived on the *London* wreck and said of the daunting conditions: 'As you start at the top it's quite green and it just gets darker and darker, the light disappears and there's loads of particles.

It's like being in a blizzard in the dark really. It's like the current is trying to pull you off the line and then you might find a nice ship comes along and stirs it all up again.[7]

Mark Beattie Edwards is the Nominated Archaeologist of the *London* and he has dived on the wreck of the *London* numerous times with Steve Ellis. 'It is quite an experience,' said Beattie. 'The visibility can be very poor – down to literally centimetres…so, generally speaking you have to dive it by feel…and sometimes when a passing ship goes nearby, the movement of water that's created means we're sometimes hanging on to the seabed and we can really see the sediment being moved around and artefacts eroding…'[8]

An Explosive Legacy of War

However, the most serious recent threat to the *London* came from something lodged in the wreck itself, which could have detonated and obliterated not only history but also taken the lives of anyone diving on it at the time.

The culprit was a German parachute mine, containing 797 kilogrammes of Hexamite explosives, which was dropped in the Thames Estuary by the Luftwaffe during the Second World War.

The mine was discovered in the summer of 2019, when Steve Ellis and his dive buddy Steve Meddle were investigating the wreck.

'He tapped me on the shoulder and drew my attention to it,' recalled Steve.

'It was lying by some iron cannons and I knew exactly what it was from pictures I'd seen in my dad's old naval history books. We must have passed over it many times but it had been there for 75 years and so I suppose if it had been going to explode then it would have done so by then.'[9]

Dives on the wreck continued, though giving the spot where the mine was located a wide berth for caution's sake.

That September bomb disposal specialists from the Royal Navy were called in, and it took eight RN mine clearance divers six days, in daunting conditions, to carefully extract the mine and then tow it out to sea. They safely detonated it, using a two-kilo charge, off Shoeburyness.

The RN team was led by Lieutenant Ben Brown, who explained how the 'complexity of the task' confronting his men was 'not be underestimated.'

Dealing with one of the largest pieces of German Second World War ordnance in the Thames Estuary presents some of the most challenging diving conditions there are to work in. With nil visibility underwater and significant tidal flow, the diving windows are extremely limited and all work on the ordnance must be done by touch.[10]

Adding to the above hazards was deteriorating weather and being close to the busy shipping channel and, as a Royal Navy source revealed, it took 'around 20 dives, to carefully lift the mine from the wreck.'

Lt Brown speculated that the mine had been dropped by a German bomber trying to inflict damage on nearby docks, though of course warships heading into or leaving Chatham also had to be on the wish list of the Luftwaffe. The mine was exactly the kind of menace that cruiser *London* dodged when sneaking out of Chatham after her rebuild concluded in March 1941. *See P71.*

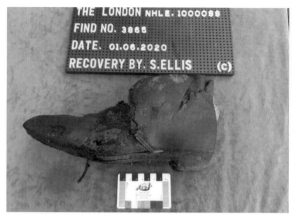 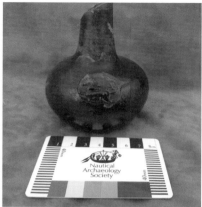

Above, left: A highly fashionable leather boot, as preserved in the wreck of the *London* for centuries and recovered by Steve Ellis in 2020. Above, right: A wine bottle recovered from the wreck of the *London*. *Photo: Courtesy Steve Ellis.*

Burying itself into the silt within the *London* wreck, the mine remained, according to Lt Brown, 'in extremely good condition given its age.'

With the wreck having survived that close call with disaster, the civilian divers of The *London* Shipwreck Trust could resume their explorations and bring more artefacts to the surface.

There are plans for a dedicated museum in Southend, where people will be able to learn about the story of the *London* lost in 1665, but whose life is now being reborn. The Save The *London* campaign is seeking to raise £200,000 initially to recover artefacts and preserve them, while also starting development of the museum proposal.

In the meantime Steve Ellis and the other diver-archaeologists will continue diving on the *London* wreck, saving her for the world, bit-by-bit. 'I love the *London* dive,' said Ellis, 'because every time we uncover new things.'[11]

For more on the The *London* Shipwreck Trust (and the Save the *London* appeal) visit the web site https://thelondonshipwrecktrust.co.uk

Notes

1. E-mail exchange between the author and Richard Endsor, who was researching and writing the forthcoming book 'The Lost Ships of Samuel Pepys's Navy'.
2. 'The London Protected, The Nore, off Southend-on-Sea, Thames Estuary, Essex', Historic England, Research Report Series no. 04-2019, by Florian Ströble, Jörn Schuster.
3. E-mail exchange between the author and Richard Endsor.
4. Quote from interview in the launch video for the Save the London appeal https://www.nauticalarchaeologysociety.org/appeal/save-the-london
5. 'The London Protected, The Nore, off Southend-on-Sea, Thames Estuary, Essex', Historic England, Research Report Series no. 04-2019, by Florian Ströble, Jörn Schuster and Research Report Series no. 15/2019, by Zoë Hazell and Emma Aitken.
6. Ibid.
7. Ibid.
8. Ibid.
9. Interview with the author, April 2021.
10. Lt Brown quotes from 'Royal Navy Detonates Huge Historic Bomb', a Royal Navy press release, 27 September 2019.
11. Interview with the author, April 2021.

SOURCES

Additional sources consulted for this new edition

Unpublished
'Journal 1944-45' of Midshipman J. B. Robathan.
'Capt J. B. Robathan, obituary,' by Peter Hore.

Books
Faulkner, Marcus, *War at Sea: A Naval Atlas 1939 – 1945*, Seaforth Publishing, 2012.
Fukuyama, Francis, *The End of History and the Last Man,* The Free Press, 1992.
Gordon, Andrew, *The Rules of the Game: Jutland and British Naval Command*, John Murray, 1997.
Hore, Peter, *Bletchley Park's Secret Source: Churchill's Wrens and the Y Service in World War II*, Greenhill Books, 2021.
Macintyre, Capt Donald, *Fighting Admiral: The Life of Admiral of the Fleet Sir James Somerville*, Evans Brothers, 1961.
Winton, John, *Sink the* Haguro: *The Last Destroyer Action of World War Two*, Pan Books, 1983.

Archives
US Naval History and Heritage Command (NHHC) -
Dictionary of American Naval Fighting Ships (DANFS) USS *Saratoga*, Ship History.

Naval-History.Net
Service histories in the Second World War of the following:
HMS *Illustrious*; HMS *London*; HMS *Valiant*.

Newspapers, journals and other publications
The Daily Telegraph, 16 April 2021.
The Sunday Telegraph, 9 May 2021.
Daily Mail, 24 February 2022.
The Guardian, 13 June 2006.
The National Interest, Summer 1989.
iNews, 28 February 2022.
The London Gazette, 21 July 1975.
Naval News, 3 March 2022.
Navy News, August 1971, June 1993.
WARSHIPS International Fleet Review, March 2010, October 2014, January 2019, June 2022

UK Parliament – Hansard
Cost of Type 23s, Written Answers, House of Commons, 5 July 2001.
Integrated Review, debate in the House of Commons, 16 March 2021.
Written questions, answers and statements, 9 March 2021, 17 March 2021.

Other

The London *Protected Wreck, The Nore, off Southend-on-Sea, Thames Estuary, Essex*: *Compositional analyses of copper alloy and pewter objects,* Historic England, Research Report Series no. 04-2019, by Florian Ströble, Jörn Schuster.

The London *Protected Wreck, The Nore, off Southend-on-Sea, Thames Estuary, Essex*: Wood identifications and recording of wooden remains recovered between 2014 and 2016, Historic England, Research Report Series no. 15/2019, by Zoë Hazell and Emma Aitken.

'Prime Minister Theresa May's speech to the Lord Mayors Banquet: 12 November 2018 – GOV.UK' – transcript.

Command Paper, *Defence in a competitive age*, published by UK Govt March 2021.

Global Britain in a competitive age, IR2021 global vision paper, published by the UK Govt March 2021.

Royal Navy press releases on City Class (Type 26) programme, 20 July 2017, 23 July 2020.

'Royal Navy Detonates Huge Historic Bomb', Royal Navy press release, 27 September 2019.

'Russia says news UK aircraft carrier "a convenient target",' 29 June 2017, BBC News online report.

'Romanian frigate drops out of NATO mission after COVID-19 outbreak', *Defense Brief*, news report, defbrief.com

Canadian Surface Combatant, Overview, Government of Canada.

Hunter Class Frigate Program, project outline, BAE Systems Australia.

'HMS Glasgow Takes a bow – Forward Section of First Type 26 is Rolled out', press release from BAE Systems, 19 April 2021.

Presidential Address to the Federal Assembly, Russia, 21 April 2021, by President Vladimir Putin. Transcript: Office of the President, Russia.

NATO MARCOM press releases: Exercise Sea Breeze, issued 21 July 2020; Exercise Sea Shield, issued 26 March 2021.

Help Save The London, fund-raising appeal launch video https://www.youtube.com/watch?v=TgJNbxRHwSg

News

Address by the President of the Russian Federation', 24 February 2022.

President Putin address to permanent members of Russia's Security Council, 3 March 2022.

BIBLIOGRAPHY

Atkinson, Rick, *Crusade,* HarperCollins, 1994

Ballantyne, Iain, *Warspite*, Pen & Sword Books Ltd., 2001

Beaver, Paul and Gander, Terry, *Modern British Military Missiles*, Patrick Stephens Ltd., 1986

Bercuson, David J. and Herwig, Holger H., *Bismarck*, Hutchinson, 2002

Blackman, Raymond V.B., *The World's Warships*, Macdonald & Co., 1969

Brodhurst, Robin, *Churchill's Anchor*, Pen & Sword Books Ltd., 2000

Brooks, Richard, *The Royal Marines*, Constable & Co. Ltd., 2002

Broome, Jack, *Make Another Signal*, Futura, 1977

Brown, Ben and Shukman, David, *All Necessary Means*, BBC Books, 1991

Brown, D.K., *Nelson to Vanguard*, Chatham Publishing, 2000
  – *Warrior to Dreadnought,* Chatham Publishing, 1997

Brownstone, David and Franck, Irene, *Timelines of War – A Chronology of Warfare from 100,000 BC to the Present,* Little, Brown & Co., 1996

Busch, Fritz-Otto, *Prinz Eugen*, Futura, 1975

Butler, Richard, *Saddam Defiant*, Phoenix, 2000

Castleden, Rodney, *British History*, Parragon, 1994

Catlow, Captain T.N., *A Sailor's Surviva*l, The Book Guild, 1999

Chant, Christopher, *Naval Forces of the World*, Collins Willow, 1984

Chesneau, Roger, *Aircraft Carriers of the World 1914 to the Present*, Brockhampton Press, 1998
  – ed., *Conway's All the World's Fighting Ships 1922-1946*, Conway Maritime Press, 1995

Claggett, John, *The U.S. Navy in Action*, Monarch, 1963

Clowes, William Laird, *The Royal Navy, a History, Vols I-VII*, Sampson Low, 1897-1903

Connell, Brian, *Knight Errant*, Hodder & Stoughton, 1955

Cowin, Hugh W., *WARSHIPS*, Penguin, 1989

Craig, Captain Chris, *Call For Fire*, John Murray, 1995

de la Billiere, General Sir Peter, *Storm Command*, HarperCollins, 1992

Dewar, Michael, *Brush Fire Wars*, Robert Hale Ltd., 1990

Dorr, Robert F., *Desert Shield*, Motorbooks International, 1991

Duncan, Andrew and Opatowski, Michel, *Trouble Spots*, Sutton Publishing, 2000

Edwards, Lieutenant Commander Kenneth, *The Grey Diplomatists,* Rich & Cowan, 1938

Elleston, Harold, *The General Against the Kremlin*, Little, Brown and Co., 1998

Fraser, Edward, *The Londons of the British Fleet*, John Lane, Bodley Head, 1908

Friedman, Norman, *Desert Victory*, Naval Institute Press, 1992

Gardiner, Robert, ed., *Conway's All the World's Fighting Ships 1947-1982, Part 1 – The Western Powers*, Conway Maritime Press, 1983

Goldrick, James, *The King's Ships Were at Sea,* United States Naval Institute, 1984

Goodman, Syd and Ballantyne, Iain, *Plymouth Warships 1900-1950*, Halsgrove, 1998

Gray, Edwyn, *Hitler's Battleships*, Pen & Sword Books Ltd., 1999

Grove, Eric, ed., *Great Battles of the Royal Navy*, Bramley Books, 1994

Hainsworth, Roger and Churches, Christine, *The Anglo-Dutch Naval Wars 1652-1674,* Sutton, 1998

Hampshire, A. Cecil, *A Short History of the Royal Navy*, Director of Public Relations (Royal Navy), 1982

Heyman, Charles, ed., *The Armed Forces of the United Kingdom 2001-2002*, Pen & Sword
  Books Ltd., 2001
Hibbert, Christopher, *Nelson A Personal History*, Penguin Books, 1995
Hill, Rear Admiral J.R., *Air Defence at Sea*, Ian Allan, 1988
  – *Anti-Submarine Warfare*, Ian Allan, 1984
  – ed., *The Oxford Illustrated History of the Royal Navy*, Oxford University Press, 1995
Hiro, Dilip, *Desert Shield to Desert Storm*, Paladin, 1992
  – *The Longest War*, Paladin, 1990
Holmes, Richard, ed., *The Oxford Companion to Military History*, Oxford University Press, 2001
Howarth, Stephen, *To Shining Sea – A History of the United States Navy 1775-1991*,

  Weidenfeld & Nicolson, 1991.
Ireland, Bernard, *Collins Jane's Warships of World War II*, HarperCollins, 1996
  – *Jane's Naval History of World War II*, HarperCollins, 1998
Ireland, Bernard and Grove, Eric, *Jane's War at Sea 1897-1997*, HarperCollins, 1997
Irving, David, *The Destruction of Convoy PQ17*, Cassell, 1968
Jane, F.T., ed., *Jane's Fighting Ships of World War I*, Studio Editions, 1990
Jones, Geoffrey, *Under Three Flags*, Corgi, 1975
Jordan, John, *Soviet Submarines 1945 to the Present*, Arms & Armour Press, 1989
  – *Soviet Warships 1945 to the Present*, Arms & Armour Press, 1992
Kemp, Paul, *British Warship Losses of the 20th Century*, Sutton, 1999
  – *Convoy!*, Cassell, 2000
Kennedy, Ludovic, *Pursuit: The Sinking of the Bismarck*, Fontana, 1975
King, Cecil, *H.M.S. (His Majesty's Ships) and Their Forbears*, The Studio Ltd., 1940
Kitson, Frank, *Prince Rupert Admiral and General-at-Sea*, Constable & Co. Ltd., 1998
Lambert, Andrew, *Battleships in Transition*, Conway, 1984
  – *The Crimean War*, Manchester University Press, 1990
Lee, Christopher, *This Sceptred Isle, 55BC-1901*, Penguin Books/BBC Books, 1998
Lenton, H.T., *British and Empire Warships of the Second World War*, Greenhill/Lionel Leventhal, 1998
Lund, Paul and Ludlam, Harry, *PQ17 Convoy to Hell*, New English Library, 1969
Macintyre, Donald, *The Battle of the Atlantic*, Severn House, 1961
Macksey, Kenneth, *The Penguin Encyclopedia of Weapons and Military Technology*, Penguin, 1995
Marriott, Leo, *Royal Navy Frigates 1945-1983*, Ian Allen, 1983
  – *Type 22*, Ian Allen, 1986
Massie, Robert K., *Dreadnought: Britain, Germany and the Coming of the Great War*, Jonathan
  Cape, 1992
Meisner, Arnold, *Desert Storm Sea War*, Motorbooks, 1991
Miller, David, *U-Boats-History, Development and Equipment 1914-1945*, Conway Maritime Press, 2000
  – *Modern Sub Hunters*, Salamander, 1992
Monsarrat, Nicholas, *The Cruel Sea*, Penguin, 1956
Moorhead, Alan, *Gallipoli*, Ballantine Books, 1993
Morriss, Roger, *Nelson*, Collins & Brown, 1996
Moynahan, Brian, *The Claws of the Bear*, Hutchinson, 1989
  – *The Russian Century*, Random House, 1994
  – *Navies in the Nuclear Age*, Conway, 1993

Neillands, Robin, *A Fighting Retreat,* Coronet, 1997

Oakley, Derek, *The Falklands Military Machine*, Spellmount, 1989

Padfield, Peter, *Maritime Supremacy and the Opening of the Western Mind*, Pimlico, 2000

Parkinson, C. Northcote, *Britannia Rules*, Sutton, 1997

Pears, Commander Randolph, *British Battleships 1892-1957*, Putnam, 1957

Peillard, Leonce, *Sink the Tirpitz!*, Granada, 1983

Penn, Geoffrey, *Fisher, Churchill and the Dardanelles*, Pen & Sword Books Ltd., 1999

Phillipson, David, *Roll on the Rodney!,* Sutton, 1999

Pope, Dudley, *Life in Nelson's Navy*, George Allen & Unwin, 1981
– *The Great Gamble,* Weidenfeld & Nicolson, 1972

Pope, Stephen and Wheal, Elizabeth-Anne, *The Macmillan Dictionary of the First World War*, Macmillan, 1997

Porten, Edward P. von der, *The German Navy in World War Two*, Pan Books, 1972

Powell, Colin, with Persico, Joseph E., *A Soldier's Way*, Hutchinson, 1995

Rayner, Caroline, ed., *Encyclopedic World Atlas*, George Philip Ltd., 1992

Ring, Jim, *We Come Unseen*, John Murray, 2001

Roberts, J.M., *Shorter Illustrated History of the World*, Helicon, 1993

Rodger, N.A.M, *The Safeguard of the Sea*, HarperCollins, 1997
– *The Wooden World - An Anatomy of the Georgian Navy*, Fontana Press, 1988

Roskill, Capt S.W., *The Navy at War 1939-1945,* Wordsworth Editions, 1998

Schofield, Carey, *Inside the Soviet Army*, Headline, 1991

Schwarzkopf, General H. Norman with Petre Peter, *It Doesn't Take a Hero*, Bantam Press, 1992

Simpson, John, *Despatches From the Barricades*, Hutchinson, 1990

Smith, Gordon, *The War at Sea – Royal & Dominion Navy Actions in World War Two,* Ian Allan, 1989

Smith, Hedrick, *The New Russians*, Vintage, 1991

Sontag, Sherry and Drew, Christopher, with Lawrence Drew, Annette, *Blind Man's Bluff,* Arrow, 2000

Southam, Brian, *Jane Austen and the Royal Navy*, Hambledon & London Ltd., 2000

Southey, Robert, *The Life of Horatio Lord Nelson*, London 1813

Steel, Nigel and Hart, Peter, *Defeat at Gallipoli,* Papermac, 1995

Sturton, Ian, ed., *All The World's Battleships*, Conway Maritime Press, 1996

Thomas , David A., *Japan's War at Sea-Pearl Harbor to the Coral Sea*, Andre Deutsch, 1978
– *Battles and Honours of the Royal Navy*, Pen & Sword Books Ltd., 1998

Thompson, Julian,*The Imperial War Museum Book of The War at Sea*, Sidgwick & Jackson, 1997

Townson, Duncan, *Dictionary of Modern History 1789-1945*, Penguin, 1995

Turner, Kara, ed., *World Handbook*, George Philip Ltd., 1995

Vat, Dan van der, *Standard of Power*, Hutchinson, 2000

Walker, Martin, *The Cold War*, Vintage, 1994

Wallis, R. Ransome, *Two Red Stripes*, Ian Allan, 1973

Warlow, Ben and Goodman, Sydney, *The Royal Navy in Focus in World War II*, Maritime Books, 1994

Watson, Bruce W., *The Changing Face of the World's Navies 1945 to Present*, Brassey's (US), 1991

Watts, Anthony, J., *The Royal Navy, An Illustrated History*, Arms & Armour, 1994

Wheal, Elizabeth-Anne and Pope, Stephen, *The MacMillan Dictionary of the Second World War*, Macmillan, 1995

Winn, Godfrey, *P.Q. 17 A Story of A Ship*, Hutchinson, 1953

Winton, John, *An Illustrated History of the Royal Navy*, Salamander, 2000
    – *Sink the Haguro!*, Pan Books, 1983
Witherow, John and Sullivan, Aidan, *War in the Gulf*, Sidgwick & Jackson 1991
Woodman, Richard, *The Sea Warriors*, Constable & Co. Ltd., 2001
Woodward, David, *The Tirpitz*, William Kimber, 1955

Other Sources of Information

Imperial War Museum Collections
Anonymous account of the Anzac Landing.
The 1914-33 Diaries of Commander C.H. Drage RN, Vol 1. 1914-1916.

The Papers of
    Cazalet, Vice Admiral Sir Peter.
    Hibbit, D.A.
    Larios, Hon. Mrs J.
    Wood, R.
    Wright, Paymaster-Commander N.

Imperial War Museum Sound Archive
Eye Witness Recordings:
    Drage, C. H. (Accession No. 006131/05)
    Hempenstall, George (Accession No. 9534/8)
    Hett, Kier (Accession No. 11325/4)
    Pearsall, John (Accession No. 9305/3)
    Priestley, Edmund (Accession No. 004641/4)
    Reynolds, Edward (Accession No. 10779/6)
    Stagles, Walter (Accession No. 4240)

National Maritime Museum
Orderbook of additional rules and orders for the better government of HMS London.

The Papers of
Broome, Captain J.
Graves, Admiral 1st Baron
Hamilton, Admiral Sir Louis Henry Keppel
Nelson, Vice Admiral Horatio

Public Record Office
Report of Proceeding of Flag Officer Commanding the First Cruiser Squadron.

Royal United Services Institution, Naval Manuscripts/National Maritime Museum
Journals kept by an anonymous lieutenant in HMS *Iron Duke, London, Kite*, and *Valiant;* and the Chinese gunboat *Eta* on passage to the East, 1874-79.

Miscellaneous Sources
A Two Day Visit to the British Forces Supporting the UN in the Former Republic of Yugoslavia 30 Nov-2 Dec 1993, account taken from the private papers of Rear Admiral T. P. McClement.
Christmas Shore Leave Cancelled For British Sailor, UK News report, filed 18 December 1995.
Documents Relating to the Naval Air Service, Volume I 1908-1918, Navy Records Society, 1969.
A Gulf Record by T.D. Elliott, 1991.
HMS London *Summary of Service*, Royal Navy Historical Branch, Revised March 1962 (S2285)
HMS London's *1946-1949 Commission Book*, an account written by Commander R. F. Leonard.
HMS London, *The Ship's First Commission, November 1963 - 1965.*
HMS London, Marine Engineering Officer's Temporary Memorandum 27/81.
HMS London *1961-1981, Paying-off Newsletter.*
HMS *London*, ship's leaflet, DPR (RN) 1991.
HMS *London* Daily Orders, 26 August 1991, 28 August 1991, 29 August 1991, 30 August 1991, 1 September 1991.
HMS *London*, Executive Officer's Temporary Memorandum, 68/91 (Lieutenant Commander A. R. Coley RN).
HMS *London*, Operation Dervish Serialized Programme, 26 August-4 September 1991.
HMS *London* 29 December 1992-28 February 1994 Programme, taken from the private papers of Rear Admiral T.P. McClement.
Letter from the Naval Historical Branch, to Mr B.D. Goodman, with reference to HMS *London*'s participation in Russian convoys, 15 January 2001.
Luncheon Menu Booklet, *Luncheon to the Ship's Company of H. M. S.* London, *19 June 1992.*
Luncheon Menu Booklet, *Luncheon to the Veterans of the Ship's Company of HMS* London *to Commemorate the 50th Anniversary of the end of World War II in the Presence of His Royal Highness the Duke of York, CVO ADC*, 8 September 1995.
Ministry of Defence, Royal Navy Press Release, 04/91, 25 March 1991, *HMS* London *Returns to Plymouth.*
Notes of Historical Prints, Relating to the 'Londons' *of the British Navy*, 1929 (Goodman Collection).
Northern Light, Official Publication of the North Russia Club, No 26, December 1991.
Speech given aboard HMS *London*, 4 February 1994, taken from the private papers of Rear Admiral T.P. McClement.
The Ceremony and Order of Service for the Re-Dedication of Her Majesty's Ship *London*, 24 March 1995.
Transcript of Colonel Gaddafi's speech at Malta, 31 March 1979, Mike North Collection.

Newspapers
Daily Mail, 2 April 1979.
Daily Telegraph, 30 March 1979, 31 March 1979, 2 April 1979.
Evening Herald, Plymouth, 13 March 1967, 6 June 1987, 8 January 1991, 17 January 1991,

6 March 1991, 7 March 1991, 21 August 1991, 27 August 1991, 28 August 1991, 29 August 1991,
30 August 1991, 2 September 1991, 4 September 1991, 9 September 1991, 16 September 1991,
17 September 1991,13 January 1993, 15 January 1993, 20 January 1993, 25 February 1993,
2 September 1993, 24 March 1995, 25 March 1995, 31 August 1995, 16 September 1996,
11 December 1997, 16 September 1998, 23 January 1999, 15 April 1999, 21 May 1999, 12 June 1999.

Malta Times, 2 April 1979.

Navy News, April 2001.

Sunday Telegraph, 10 June 2001.

The News, Portsmouth, 8 December 1961, 31 October 1963, 12 November 1963,
22 November 1963, 23 November 1963, 25 November 1963, 5 February 1964, 20 May 1964,
30 June 1964, 2 July 1964, 22 August 1964, 25 August 1964, 26 August 1964, 31 August 1964,
16 March 1966, 27 April 1966, 18 May 1966, 21 May 1966, 24 October 1966, 30 November 1966,
8 March 1967, 13 March 1967, 5 July 1967, 29 September 1967, 12 March 1968,
19 March 1968, 6 April 1968, 11 October 1968, 18 October 1968, 27 June 1970, 10 July 1970,
6 August 1970, 4 January 1971, 8 October 1971, 24 April 1972, 10 June 1976, 26 April 1980,
16 September 1981, 9 December 1981, 11 December 1981, 20 January 1982, 23 March 1982,
23 April 1982, 9 June 1982, 2 May 1987.

Western Morning News, 21 August 1965, 24 June 1987, 26 March 1991, 2 September 1991,
25 March 1995.

Magazines & Other Publications

BAE SYSTEMS Response, April/May 2002.

DML Link, August 1995, article by the author on HMS *London*'s 'Thursday War'.

Marine News, Vol 12, 1958.

Navy International, October 1991.

Periscope, the newspaper of Chatham Royal Dockyard, April 1980.

WARSHIPS International Fleet Review, Summer '98, Autumn '98, Oct/Nov 2000,
Oct/Nov 2001, June/July 2002.

VJ DAY, A Special Publication to Mark the end of World War II, Western Morning News Co.,
1995, written by Iain Ballantyne and Michael Charleston.

Witness to War, Los Angeles Times, 1991.

Unpublished Documents

Bruty, Gordon, *HMS London 1940-1946.*

Dale, Frederick, *The Memoirs of Frederick Dale, Master at Arms, HMS London October 1965 –*
May 1968 & April 1970 – August 1972.

Harwood, Frederick, *The Memoirs & Papers of Bandmaster Frederick Harwood.*

Parker-Jervis, Christopher, *H.M.S. London and the Yangtse Affair-April, 1949.*

INDEX